Annals of the ICRP

ICRP PUBLICATION 92

Relative Biological Effectiveness (RBE), Quality Factor (Q), and Radiation Weighting Factor (w_R)

Editor
J. VALENTIN

PUBLISHED FOR

The International Commission on Radiological Protection

by

Los Angeles | London | New Delhi
Singapore | Washington DC

CONTENTS

ICRP Publication 92

Guest Editorial

A CURRENT VIEW ON RADIATION WEIGHTING FACTORS AND EFFECTIVE DOSE

The International Commission on Radiological Protection (ICRP) developed radiation weighting factors (w_R) for use in radiological protection in *Publication 60* (ICRP, 1991), and Table 1 includes the w_R values that were recommended. Since 1990, there have been substantial developments in biological and dosimetric knowledge that justify a re-appraisal of w_R values and how they may be derived. This re-appraisal is the principal objective of the present report (*Publication 92*), which was a joint venture between ICRP Committees 1 and 2.

This report is one of a set of documents being developed by ICRP Committees in order to advise the ICRP on the formulation of its next Recommendations for Radiological Protection. Here, we summarise our personal views on the principal implications of the report and how the ICRP might proceed with the derivation of w_R values ahead of its forthcoming recommendations. Table 1 provides a comparison of w_R values from *Publication 60* with values proposed in the present report.

In *Publication 60*, the ICRP defined effective dose as the doubly weighted sum of absorbed dose in all the organs and tissues of the body. Dose limits are set in terms

Table 1. A comparison of existing w_R values and those proposed to the ICRP

Type and energy range of incident radiation		Radiation weighting factor (w_R)	
		Publication 60	Proposed[c]
Photons, all energies		1	1
Electrons and muons (all energies)[a]		1	1
Protons (incident)		5	2
Neutrons, energy	< 10 keV	5	
	10 keV–100 keV	10	Use the proposed w_R function in Fig. 1 below
	> 100 keV–2 MeV	20	
	> 2 MeV–20 MeV	10	
	> 20 MeV	5	
Alpha particles, fission fragments, and heavy ions[b]		20	20[d]

[a] Exclude Auger electrons from emitters localising to cell nucleus/DNA- special treatment needed.

[b] Use Q-LET relationships of *Publication 60* for unspecified particles.

[c] Changes for neutron energies < 1 MeV are required to account for gamma contribution to internal organs (see text).

[d] ICRP Committee 4 Task Group on Radiological Protection in Space Flight to consider w_R for high energy neutrons and heavy ions of LET > 200 keV/μm.

of effective dose and apply to the individual for radiological protection purposes, including the assessment of risk in general terms.

The values of both radiation and tissue weighting factors depend on current knowledge of biophysics and radiobiology and, accordingly, the ICRP acknowledges that judgements on these factors may change from time to time. When such changes are made, the ICRP does not recommend that any attempt is made to correct individual historical estimates of effective dose or of equivalent dose (in a single tissue or organ) that have been incurred.

We believe that the ICRP should continue the use of w_R values that relate, for external radiation, to the incident field. For radionuclide intakes, w_R values should relate to the internal fields that cause the doses to specific organs and tissues. These w_R values are intended to take account of the radiation quality component of effective dose.

For calculating effective dose, we suggest the use of the same set of w_R values for all organs/tissues. This is a judgement of practicability not based on firm radiobiological knowledge.

For photons and beta particles, there are cellular and biophysical data that indicate energy-dependent variations of up to a few-fold in relative biological effectiveness (RBE) at low doses; the extent to which this variation applies to cancer risk is not clear, and available epidemiological data suggest little variation. Also, given the UNSCEAR 2000 judgement of several-fold uncertainty on the judgement of the nominal risk coefficient for cancer, including that for the dose and dose-rate effectiveness factor (DDREF), we do not see the need to ascribe different values of w_R to different low-linear-energy-transfer (LET) radiations. A w_R of 1 may therefore be retained for all low-LET radiations.

Publication 60 had ascribed a w_R of 5 to all protons of energy >2 MeV. We suggest that this is a significant overestimate of the biological effectiveness of these protons and, for incident protons of practical importance (>10 MeV), we propose a w_R of 2, i.e. a value which is somewhat in excess of the w_R for other low LET radiations in order to account for the densely ionising secondaries released by high energy protons in the human body. A joint ICRU–ICRP Task Group on Doses from Cosmic Ray Exposure to Aircrew is expected to provide additional information in this respect.

Auger emitters, which have the potential to localise to the cell nucleus and bind to DNA, were recognised in *Publication 60* as a special case for low-LET radiation. We believe that such Auger emitters will need continued special attention in radiological protection. It would, however, be necessary to obtain specific physiological and biophysical data on the uptake, cellular localisation, and turnover of candidate Auger-emitting compounds in order to ascribe meaningful w_R values. This would involve considerable effort.

For neutrons, we suggest that the ICRP continues the use of w_R values that depend upon the energy of the incident neutrons. However, a continuous function (see Fig. 1) should be used, rather than the step function given in *Publication 60*. This procedure will reduce problems of computation of effective dose but should not be taken to imply precise knowledge of the underlying biological effectiveness. The

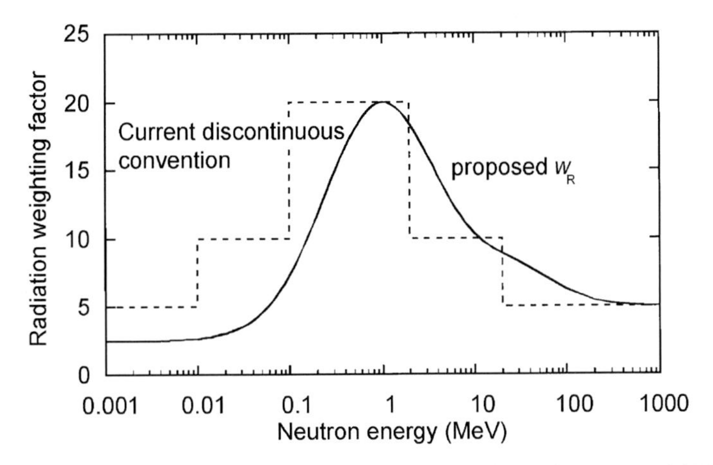

Fig. 1. The radiation weighting factor w_R for neutrons introduced in *Publication 60* (ICRP, 1991) as a discontinuous function of the neutron energy (- - -) and the proposed modification (—).

neutron w_R should be decreased for neutron energies below around 1 MeV to take account of the absorbed dose contribution by low-LET γ rays ($w_R = 1$) that are induced in the body by neutron capture. This γ component is considerably greater than implied in *Publication 60*, and the relative contribution of the densely ionising dose component is correspondingly smaller. The modified values of w_R, including those at high neutron energies, are—unlike the earlier values—consistent with an internal weighting factor that depends on LET, which facilitates the comparison of measurements with computed values of effective dose.

We recognise that there are uncertainties in ascribing appropriate w_R values for neutrons of high energy (> 20 MeV). This is a significant issue primarily for exposures at high altitude. We anticipate that the current ICRP Committee 4 Task Group on Radiological Protection in Space Flight will advise on these matters.

For alpha particles and for all heavy ions, we suggest that a w_R of 20 continues to be appropriate, but recognise considerable remaining uncertainty with respect to heavy ions of LET greater than around 200–300 keV/μm. We anticipate that the uncertainty will be considered in depth by the Task Group on Space Flight. For specific circumstances involving heavy charged particles, we suggest that the quality factor and the numerical $Q(L)$ values introduced in *Publication 60* be used for deriving w_R.

Measurements are an essential element of radiological protection, e.g. determinations of radionuclide intakes, ambient dose equivalent, and personal dose equivalent in a defined phantom. The reference quantities and the measurement procedures are generally chosen to provide conservative estimates of effective dose. The intention is to ensure that compliance with measured quantities may be used to demonstrate compliance with legal limits. Estimates of effective dose that are close to or above these limits should prompt follow-up computations specific to the individual. Dosimetric anomalies, as may occur in highly non-uniform external fields or with the intake of radionuclides, may require assessments that take individual characteristics of the exposed person and specifics of the exposure situation into account.

Finally, it is important to remember that effective dose is a quantity intended for use in radiological protection and was not developed for use in epidemiological studies or other specific investigations of human exposure. For these other studies, absorbed dose in the organs of interest and specific data relating to the RBE of the radiation type in question are the most relevant quantities to use.

Roger Cox
Albrecht M. Kellerer

PREFACE

In 1998, on the proposal of its Committee 1 on Radiation Effects, the International Commission on Radiological Protection (ICRP) established a Task Group on Radiation Quality Effects in Radiological Protection.

The Terms of Reference of the Task Group were to collate and evaluate data for alpha particles, neutrons, and protons; to consider deterministic and stochastic effects; and to provide comments on effects after both acute and prolonged exposure.

The Terms of Reference also requested the Task Group to examine the methods of handling differences and uncertainties in radiation quality effects for the purposes of radiological protection.

The Task Group had the following full members:

R.J.M. Fry	A.M. Kellerer	G. Dietze

The Task Group was chaired by Dr Fry from 1998 to April 2001, and by Professor Kellerer from May 2001 until completion of the report in the spring of 2003.

The corresponding members of this Task Group were:

D. Goodhead	W.K. Sinclair	H.R. Withers
A.A. Edwards	M. Harms-Ringdahl	P. Pihet

The extensive contribution of Dr Sinclair was particularly important for this project.

During the period of preparation of this report, the membership of ICRP Committee 1 was:

(1997–2001)

R. Cox (Chairman)	A.V. Akleyev	R.J.M. Fry
J.H. Hendry	A.M. Kellerer	C.E. Land
J.B. Little	K. Mabuchi	R. Masse
C.R. Muirhead (Secretary)	R.J. Preston	K. Sankaranarayanan
R.E. Shore	C. Streffer	R. Ullrich
K. Wei	H.R. Withers	(1998–; Vice-Chairman)

(2001–2005)

R. Cox (Chairman)	A.V. Akleyev	M. Blettner
J.H. Hendry	A.M. Kellerer	C.E. Land
J.B. Little	C.R. Muirhead (Secretary)	O. Niwa
D. Preston	R.J. Preston	E. Ron
K. Sankaranarayanan	R.E. Shore	F.A. Stewart
M. Tirmarche	R. Ullrich (Vice-Chairman)	P.-K. Zhou

The report was approved by the Commission in October 2002. The Commission approved of the guest editorial and the executive summary in January 2003.

Relative biological effectiveness (RBE), quality factor (Q), and radiation weighting factor (w_R)

ICRP Publication 92

Approved by the Commission in January 2003

Abstract–The effect of ionising radiation is influenced by the dose, the dose rate, and the quality of the radiation. Before 1990, dose-equivalent quantities were defined in terms of a quality factor, $Q(L)$, that was applied to the absorbed dose at a point in order to take into account the differences in the effects of different types of radiation. In its 1990 recommendations, the ICRP introduced a modified concept. For radiological protection purposes, the absorbed dose is averaged over an organ or tissue, T, and this absorbed dose average is weighted for the radiation quality in terms of the radiation weighting factor, w_R, for the type and energy of radiation incident on the body. The resulting weighted dose is designated as the organ- or tissue-equivalent dose, H_T. The sum of the organ-equivalent doses weighted by the ICRP organ-weighting factors, w_T, is termed the effective dose, E. Measurements can be performed in terms of the operational quantities, ambient dose equivalent, and personal dose equivalent. These quantities continue to be defined in terms of the absorbed dose at the reference point weighted by $Q(L)$.

The values for w_R and $Q(L)$ in the 1990 recommendations were based on a review of the biological and other information available, but the underlying relative biological effectiveness (RBE) values and the choice of w_R values were not elaborated in detail. Since 1990, there have been substantial developments in biological and dosimetric knowledge that justify a re-appraisal of w_R values and how they may be derived.

This re-appraisal is the principal objective of the present report. The report discusses in some detail the values of RBE with regard to stochastic effects, which are central to the selection of w_R and $Q(L)$. Those factors and the dose-equivalent quantities are restricted to the dose range of interest to radiation protection, i.e. to the general magnitude of the dose limits. In special circumstances where one deals with higher doses that can cause deterministic effects, the relevant RBE values are applied to obtain a weighted dose. The question of RBE values for deterministic effects and how they should be used is also treated in the report, but it is an issue that will demand further investigations.

This report is one of a set of documents being developed by ICRP Committees in order to advise the ICRP on the formulation of its next Recommendations for Radiological Protection.

Thus, while the report suggests some future modifications, the w_R values given in the 1990 recommendations are still valid at this time. The report provides a scientific background and suggests how the ICRP might proceed with the derivation of w_R values ahead of its forthcoming recommendations.

Keywords: Equivalent dose; Effective dose; Radiation protection; Stochastic; Deterministic.

1. INTRODUCTION

(1) Unambiguous definition of the basic quantities is the precondition of a sound system of radiation dosimetry and radiation protection. Before the 1990 recommendations of the International Commission on Radiological Protection (ICRP, 1991), all dose-equivalent quantities were defined in terms of a weighting factor, the quality factor, $Q(L)$, that was applied to the absorbed dose at a point. The weighted absorbed dose was called the dose equivalent, H. Averaging over an organ or a tissue, T, provided the mean organ or tissue dose equivalents, H_T, and their organ weighted average was the effective dose equivalent, H_E:

$$H_T = \int_m \int_L Q(L) D_L \, dL dm/m \quad \text{and} \qquad H_E = \Sigma_T w_T H_T \tag{1.1}$$

where D_L is the distribution of absorbed dose in unrestricted linear energy transfer (LET), and the integral ranges over LET and the mass, m, of the organ.

(2) $Q(L)$ is a function of unrestricted LET specified by the International Commission on Radiation Units and Measurements (ICRU, 1980) as:

$$L_\infty = dE/dx \tag{1.2}$$

where dE is the energy lost by a charged particle in traversing a distance dx. The unrestricted LET, L_∞, is commonly denoted by L (ICRU, 1980, 1993b).

(3) The unit of absorbed dose and of all dose-equivalent quantities is the joule per kilogramme. To avoid confusion between the quantities, the term 'gray (Gy)' is used for this unit when reference is made to absorbed dose, while the term 'sievert (Sv)' is used with the dose-equivalent quantities.

(4) The 1990 recommendations (ICRP, 1991) introduced a modified concept to take into account the differences in the effects of different types of radiation. For radiation protection purposes, the absorbed dose is averaged over an organ or tissue, T, and this absorbed dose average is weighted for the radiation quality in terms of the radiation weighting factor, w_R, for the type and energy of radiation incident on the body. The resulting weighted dose was designated as the organ- or tissue-equivalent dose:

$$H_T = \Sigma_R w_R D_{T,R} \tag{1.3}$$

where $D_{T,R}$ is the absorbed dose averaged over the tissue or organ, T, due to the incident radiation, R.

(5) The sum of the organ-equivalent doses weighted by the ICRP organ weighting factors, w_T, is termed the effective dose, E:

$$E = \Sigma_T w_T H_T \tag{1.4}$$

(6) The reason for replacing the quality factor, i.e. the *Q–L* relationship, with w_R values in the definition of the organ-equivalent doses and the effective dose was that the Commission believed:

> 'that the detail and precision inherent in using a formal *Q–L* relationship to modify absorbed dose to reflect the higher probability of detriment resulting from exposure to radiation components with high LET is not justified because of the uncertainties in the radiological information'.

(7) Measurements are performed in terms of the operational quantities, ambient dose equivalent, and personal dose equivalent. These quantities are still defined in terms of the absorbed dose at the reference point weighted by $Q(L)$.

(8) In the same recommendations (ICRP, 1991) that introduced w_R and E, the Commission modified the earlier relationship (ICRP, 1977) between $Q(L)$ and L (see Fig.1.1). The change reflected higher relative biological effectiveness (RBE) values for intermediate-energy neutrons, and the reduced effectiveness of heavy ions with L greater than 100 keV/μm, as suggested by the Joint ICRU and ICRP Task Group on 'The Quality Factor in Radiation Protection' (ICRU, 1986). The current *Q–L* relationship is compared with the earlier convention in Fig. 1.1 and it is specified in Table 1.1.

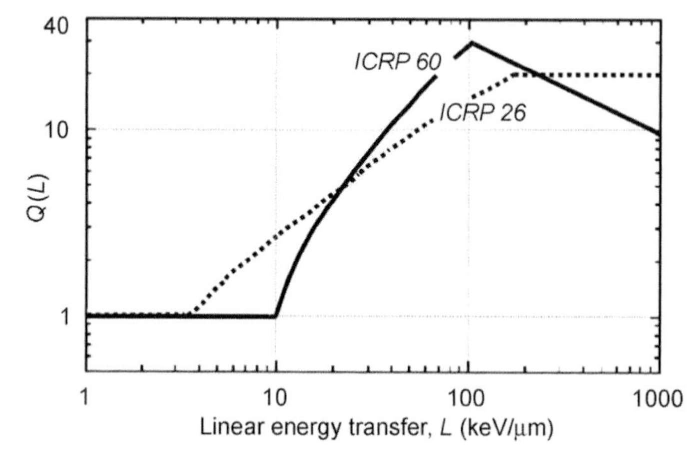

Fig. 1.1. The earlier convention for the quality factor, $Q(L)$, as a function of linear energy transfer according to *Publication 26* (------: ICRP, 1977) and the current convention according to *Publication 60* (———: ICRP, 1991).

Table 1.1. Quality factor relationship

Unrestricted linear energy transfer, L (keV/μm)	Quality factor, $Q(L)$
<10	1
10—100	$0.32\,L-2.2$
>100	$300/\sqrt{L}$

(9) The Q–L relationship, the Commission contended:

'was originally intended to do no more than provide a rough indication of the variation of the values of Q with changes of radiation, but it was often interpreted to imply a spurious precision which the Commission hopes will not be inferred from the new radiation weighting factors'.

(10) The Commission selected the values for w_R shown in Table 1.2 'based on a review of the biological information, a variety of exposure circumstances and inspection of the traditional calculations of the ambient dose equivalent'. The 1991 report did not elaborate which RBE values had been considered nor how a single value of w_R for each radiation category had been chosen. However, as has been stated above, the modified convention for $Q(L)$ reflected the advice of the ICRU-ICRP Liaison Committee (ICRU, 1986), and the selected values of w_R (Table 1.1) were chosen to be broadly compatible with $Q(L)$. The numerical inter-relationship between w_R and $Q(L)$ is assessed in detail in Chapter 4.

(11) ICRU (1986) called the quality factor Q and stated

'The dimensionless factor Q, is termed the quality factor. The selection of its numerical values depends not only on appropriate biological data, but also on judgement. Judgement may include deciding which biological endpoints are of importance and how their RBE values should be weighted in order to establish Q. It should also include an assumption about the shape of the dose-effect relationship for human risk at low doses. Linearity is usually assumed in this range, but conditions such as dose rate may have to be taken into account in determining the slope. The importance of Q derives from the fact that it establishes the values of the absorbed dose of any radiation that engenders the same risk as a given absorbed dose of a reference radiation'.

(12) As shown in Table 1.2, a w_R of 1 was selected for all low-LET radiations, i.e. x and γ rays of all energies as well as electrons and muons.

A smooth curve, considered an approximation, was fitted to the w_R values as a function of incident neutron energy in terms of the relationship:

Table 1.2. Radiation weighting factors (ICRP, 1991)

Radiation type and energy range		Radiation weighting factor, w_R
Photons, all energies		1
Electrons and muons, all energies		1
Neutrons, energy	$<$ 10 keV	5
	10–100 keV	10
	$>$ 100 keV–2 MeV	20
	$>$ 2–20 MeV	10
	$>$ 20 MeV	5
Protons, other than recoil protons, energy $>$ 2 MeV		5
α particles, fission fragments, heavy nuclei		20

$$w_R = 5 + 17\exp\left(-(\ln(2E_n))^2/6\right) \qquad (1.5)$$

where E_n is the neutron energy in MeV. This relationship is purely calculational.

(13) No specifications were given to derive the equivalent dose or the effective dose from Auger electrons. The assessment of their relative effectiveness will have to be based on microdosimetry; no attempt to do so was made in *Publication 60* (ICRP, 1991).

(14) For types of radiation and energy not included in Table 1.1, the ICRP suggests that w_R be obtained by calculation of the ambient quality factor, i.e. the mean $Q(L)$ at a depth of 10 mm in the ICRU sphere exposed to the aligned radiation field (ICRU, 1985):

$$q^* = \int Q(L)D_L \mathrm{d}L/D \qquad (1.6)$$

where D_L dL is the absorbed dose at 10 mm depth between L and $L+\mathrm{d}L$. The quality of radiation can also be defined in terms of lineal energy, y (ICRU, 1986).

(15) This report is concerned with the current use of w_R and $Q(L)$ for the derivation of equivalent doses for radiation protection related to stochastic effects. As explained in Chapter 4, w_R is a simplified concept to replace $Q(L)$ in practice. RBE values with regard to stochastic effects are central to the selection of w_R and $Q(L)$, and these factors and the dose-equivalent quantities are restricted to the dose range of interest to radiation protection, i.e. to the general magnitude of the dose limits. In special circumstances where one deals with higher doses that can cause deterministic effects, the relevant RBE values are applied to obtain a weighted dose that takes the relative effectiveness of different types of radiation into account. The question of RBE values for deterministic effects and how they should be used is treated separately in Chapter 5.

2. THE CONCEPT OF RBE, RELATIVE BIOLOGICAL EFFECTIVENESS

2.1. Background

(16) The effect of ionising radiation is influenced by the dose, dose rate, and quality of the radiation. In 1931, Failla and Henshaw reported on determination of the relative biological effectiveness (RBE) of x rays and γ rays. This appears to be the first use of the term 'RBE'. The authors noted that RBE was dependent on the experimental system being studied. Somewhat later, it was pointed out by Zirkle et al. (1952) that the biological effectiveness depends on the spatial distribution of the energy imparted and the density of ionisations per unit path length of the ionising particles. Zirkle et al. coined the term 'linear energy transfer (LET)' to be used in radiobiology for the stopping power, i.e. the energy loss per unit path length of a charged particle.

(17) RBE has been used in somewhat different ways in radiobiology and radiation protection. In the former, it equals the ratio of the absorbed doses of two types of radiation that produce the same specified effect. In radiation protection, a more general parameter is required as a weighting factor for absorbed doses of radiation of different qualities to enable comparison and addition. Initially, the ICRP employed the term 'relative biological efficiency' for this purpose, when it based its recommendations (ICRP, 1951) 'on considerations of the equivalent energy absorbed in tissue coupled with the appropriate relative biological efficiency'.

(18) However, it was soon realised that the use of RBE as a weighting factor is complicated by the fact that it is dependent on dose, dose rate, fractionation, and the cells or tissues in which the effect is being assessed. Consequently, in 1959, the ICRU recommended that the term:

> 'RBE be used in radiobiology only and that another name be used for the linear-energy-transfer dependent factor by which absorbed doses are to be multiplied to obtain for purposes of radiation protection a quantity that expresses on a common scale for all ionising radiations the irradiation incurred by exposed persons'.

The name chosen for this factor was quality factor (QF) and the dose equivalent (DE) was defined as the product of absorbed dose D and QF. These recommendations were taken up and were endorsed by the RBE Committee (ICRU–ICRP, 1963). Currently, RBE is only used in radiation protection in terms of the derived quantities, quality factor, $Q(L)$, and radiation weighting factor, w_R.

(19) RBE has been criticised in some cases because it was thought to have been inappropriately employed to explore the biological mechanisms of radiation action. However, until the biological effectiveness of different types of ionising radiations can be accounted for in terms of individual contributing factors, the summary concept of RBE will continue to be used (see Sinclair, 1996).

(20) In the case of radiation protection in which RBE values for different types of radiation for the induction of stochastic effects are used to generate values of w_R and Q, there are a number of concerns:

- for the ratio of doses to be meaningful for interpretation of the relative effectiveness of the radiation, it is necessary to know the characteristics of the radiations at the target cells or tissues. Usually, however, RBE values are related to the characteristics of the incident radiation that is appropriate if the quality of radiation under study is not substantially altered in the passage to tissues deep in the body. In cell studies and animal experiments, this condition is usually met. In the exposure of the human body to neutrons, the external and internal field characteristics differ markedly; this issue is discussed in detail with regard to w_R values for neutrons in Chapter 4.
- there is an important assumption in the use of RBE, namely that the effects of various types of radiation differ quantitatively and not qualitatively. Recent studies on the nature of DNA damage, including mutations, especially in the case of heavy charged particles, have shown differences from the changes induced by low-LET radiation. The question is whether differences such as in the induction of complex DNA lesions that are recalcitrant to repair make the assumption untenable. If the endpoint under study that is induced by the different types of radiation is precisely the same, e.g. the severity (degree of malignancy) of radiogenic cancers is not influenced by radiation quality, the assumption that any differences due to radiation quality are only quantitative should be valid.
- for control purposes in radiation protection, a single and maximum value is required, defined by the ICRP–ICRU RBE Committee (ICRU–ICRP, 1963) as the RBE at minimal doses. This value, RBE_M, is determined as the ratio of the initial slopes of the dose–effect curves for the radiation under study and the reference radiation. Determination of the initial linear slopes of the dose–response curves for the induction of specific cancers by different radiations is not a trivial task. Based on the assumption that the linear-quadratic model is applicable to the dose–response relationship of the low-LET reference radiation, the initial slope can be determined from the response at a low dose rate. The RBE values quoted in 'The Quality Factor in Radiation Protection' (ICRU, 1986) are based on 'different types of exposure, for example single, fractionated, and low-dose-rate γ radiation and x rays'.

(21) The use of RBE_M was introduced to avoid the problems with RBE values which were determined at a single dose level not on the initial linear component of the dose–response curve for the reference radiation. However, the RBE_M approach is obviously not applicable if—as may be the case for certain types of bone tumours and for some skin tumours—there is a threshold, or if the response is best described by a dose-squared model. Such dose responses present a problem in how the effectiveness of high-LET radiations should be estimated and underline the need to develop a method of assessing the effect directly *without comparison to a reference radiation.* This would entail the development of an acceptable method of extrapolation across species.

2.2. Reference radiation

(22) The original reference radiation for the weighting factor in radiation protection was stated to be γ rays from radium (ICRP, 1951; Taylor, 1984):

'The relative biological efficiency of any given radiation has been defined by comparison with the γ radiation from radium filtered by 0.5 mm of platinum. It has been expressed numerically as the inverse of the ratio of the doses of the two radiations required to produce the same biological effect under the same conditions. It has been assumed, for purposes of calculations, that the relative biological efficiency of a given radiation is the same for all effects mentioned in the Introduction with the single exception of gene mutations. The following values have been adopted:

Radiation	Relative biological efficiency
γ rays from radium	
x rays of energy 0.1–3 MeV	1
β rays	
Protons	10
Fast neutrons <20 MeV	10
α rays	20

The effective figure for slow neutrons should be derived in any given case from an evaluation of the separate contributions to the biological effect by protons arising from the disintegration of the nitrogen nuclei and by γ rays arising from capture of neutrons by hydrogen nuclei'.

(23) Apart from the notable fact that hereditary effects, the 'gene mutations' were excluded, the weighting factor was understood to be of a very general nature, as the list of 'effects mentioned in the Introduction' indicates:

- superficial injuries;
- general effects on the body, particularly the blood and blood-forming organs, e.g. production of anaemia and leukaemias;
- the induction of malignant tumours;
- other deleterious effects including cataract, obesity, impaired fertility, and reduction of life span;
- genetic effects.

(24) While γ rays were declared to be the reference radiation, the above table shows that a value of 1 was, in fact, assigned to a broad range of photon and electron radiations. It was, thus, no major reversal when ICRP changed its position in 1955 and stated:

'RBE should be expressed in terms of the pertinent biological effectiveness of ordinary x rays taken as 1 (average specific ionisation of 100 ion pairs per μm of water or linear energy transfer of 3.5 keV per μm of water)'.

(25) Subsequently, there have been different conventions, but the issue became less critical after the concept of RBE was clearly separated from the adopted weighting factors, $Q(L)$ and later w_R.

(26) With regard to the weighting factors, there is no specific reference radiation. Instead, a value of 1 is assigned to a range of low-LET radiations. In *Publication 60* (ICRP, 1991), w_R was set at 1 for all photons and electrons, which suggests that the differences of effectiveness between different photon radiations are not considered to be of sufficient consequence to require explicit accounting in radiation protection regulations. The current convention for $Q(L)$ is in line with the recommendation for w_R, because it attributes a value of 1 to $Q(L)$ for unrestricted LET, $L < 10$ keV/μm. Since LET values in excess of 10 keV/μm are reached only by electrons below 1.5 keV, it is clear that the convention for $Q(L)$ is largely consistent with $w_R = 1$ for x rays and electron radiations.

(27) For RBE—which serves as important input into the selection of the weighting factor—the situation is different. It is essential that any statement of RBE must be accompanied by a specification of the reference radiation. The significant difference (see Fig. 2.3) between the (dose-average) LET of ^{60}Co (about 0.4 keV/μm) or ^{137}Cs γ rays (about 0.8 keV/μm), and that of 200 kV x rays (about 3.5 keV/μm) makes it clear that RBE values can differ substantially depending on which photon radiation is taken as reference.

(28) While there is no need for an exclusive convention, it is nevertheless convenient to adopt a reference radiation that is understood to apply whenever there is no explicit statement to the contrary. There are practical arguments to favour γ rays for this purpose. It is difficult and expensive to determine the initial slope of dose responses of the induction of cancer in animals, especially with low-dose-rate x rays rather than low-dose-rate γ rays. For this and a number of other reasons, hard γ rays are preferable as the reference radiation because:

- most experimental animal studies of cancer induction and life shortening (and deterministic effects) have been carried out with γ rays, and, importantly, some with exposures at low dose rates;
- the most important body of data for estimating radiogenic cancers in humans are from the atomic bomb survivors who were exposed to γ rays;
- hard γ rays have the lowest LET (dose average LET, L_D, 0.4 keV/μm or less) among photon radiations;
- the distribution of the deposition of energy from γ rays in large fields is more uniform than with x rays.

(29) The convention to assign the same w_R to photons and electrons of all energies is advantageous for the purposes of radiation protection, since it facilitates measurements as well as calculations. It is, of course, restricted to the w_R as part of the definition of effective dose and equivalent doses, and it does not imply that all low-LET radiations are assumed to be equally effective. For risk assessment, for example in a comparison of γ rays and x rays or conventional x rays and soft x rays, the w_R value is not applicable, LET or microdosimetric parameters must be invoked, and radiobiological or radio-epidemiological data need to be used. The general magnitude of RBE values that are expected and have been observed with photons or electrons will, therefore, be reviewed briefly.

2.2.1. In-vitro studies

(30) It has long been recognised that, especially at low doses, low-LET radiations do not all have the same effectiveness. Conventional 200 kV x rays are considered to be about twice as effective at low doses as high-energy γ rays based upon studies of mutations in *Tradescantia*, aberrations in human lymphocytes, and killing of mouse oocytes (Bond et al., 1978). Fast electrons may be even less effective than γ rays. These differences are of interest in themselves, but must also be taken into account when different photon radiations are used as reference radiation (Sinclair, 1985; ICRU, 1986).

(31) The most reliable and detailed data on photon RBE exist for chromosome aberrations in human lymphocytes. In choosing the $Q(L)$ values, the report of the Joint ICRP and ICRU Task Group (ICRU, 1986) has given special consideration to observations on chromosome aberrations in human lymphocytes (Edwards et al., 1982) for 15 MeV electrons, ^{60}Co γ rays, and 250 kV x rays. These authors obtained the data for dicentrics in human lymphocytes listed in Table 2.1, i.e. they have shown substantial differences of effectiveness for different types of penetrating low-LET radiations. New data have since confirmed and extended these findings.

(32) Sasaki et al. determined the yields of dicentrics over a broad range of photon energies. Figure 2.1 gives the linear coefficients (and standard errors) from linear-quadratic fits to the dose dependencies. The upper panel refers to peripheral human lymphocytes (Sasaki et al., 1989; Sasaki, 1991), and the lower panel gives data for the cultured cell line m5S from embryonic mouse cells (Sasaki, 2003, private communication); these cells are immortalised but not malignantly transformed and they retain near-diploid chromosome constitution. The circles relate to γ rays and broad x-ray spectra, and the squares relate to characteristic x rays and mono-energetic photons from synchrotron radiation.

(33) Figure 2.1 demonstrates that there is a substantial decrease of the yield of dicentrics from conventional x rays to γ rays for both human lymphocytes and mouse cells. Photon energies below 20 keV are of particular interest with regard to biophysical considerations, but are less relevant to exposure situations in radiation protection. They are included here to show the full trend of energy dependence.

(34) Figure 2.2 represents analogous data obtained by Schmid et al. (2002b) for chromosome abberations in human lymphocytes from blood samples of one and the same donor. The upper and lower panels give the initial slope for dicentric and acentric fragments, respectively. For dicentrics, it is seen that the RBE_M of moderately filtered 200 kV x rays is about 2–3 relative to γ rays, while the RBE_M of

Table 2.1. Low dose coefficient (and standard error) for the induction of dicentrics in human lymphocytes by low-linear-energy-transfer penetrating radiation (Edwards et al., 1982)

Radiation type	Dicentrics per cell/Gy
15 MeV electrons	0.0055 (±0.0112)
^{60}Co γ rays	0.0157 (±0.0029)
250 kV x rays	0.0476 (±0.0054)

mammography x rays (29 kV) relative to the moderately filtered 200 kV x rays is somewhat in excess of 1.5. The corresponding values for acentrics are about 2 for comparison of 200 kV x rays and γ rays, and less than 1.5 for comparison of mammography x rays and moderately filtered 200 kV x rays.

(35) For the RBE_M of mammography x rays relative to γ rays, a value of about 5 is obtained with regard to dicentrics, and a value of about 3 with regard to acentrics.

(36) The difference by a factor of 2–3 in the low dose effectiveness of conventional x rays and γ rays is known and, even if this should apply equally to radiation-induced late effects, this would not necessarily require a departure from the current convention which assigns $w_R = 1$ to all photon radiations. However, the difference needs to be noted whenever risk estimates are derived from exposures to γ rays and are then applied to x rays. As stated, the difference is also important when RBE_M values are derived for densely ionising radiation, and different low-LET reference radiations are chosen.

(37) Apart from these considerations, it is uncertain whether the marked dependence of RBE_M on photon energy for chromosome aberrations is also representative for late radiation effects in man. The dependence of RBE_M on photon energy for dicentric chromosomes reflects the fact that the dose dependencies have a large

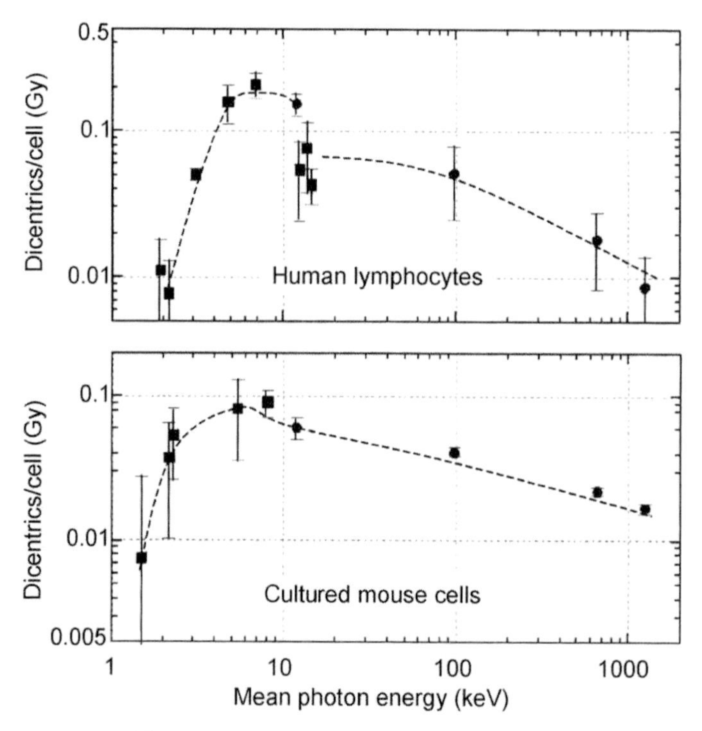

Fig. 2.1. Photon energy and linear coefficients (and standard errors) of the dose dependence for dicentric chromosomes in human peripheral blood lymphocytes (upper panel) (Sasaki et al., 1989; Sasaki, 1991) and in mouse m5S cells (lower panel) (Sasaki, 2003, private communication). Squares, mono-energetic photons; circles, broad x-ray spectra and high energy γ rays. The broken curves are inserted for visual guidance.

curvature for ^{60}Co γ rays [$\alpha/\beta = 0.2$ Gy in the data reported by Schmid et al. (2002b)] but little curvature for 29 kV x rays ($\alpha/\beta = 1.9$ Gy). If there were no curvature below 1 Gy in the dose relationships for chromosome aberrations, the RBE_M of the 29 kV x rays would be only 1.65 against the ^{60}Co γ rays. Since the dose dependence for solid tumours among the atomic bomb survivors indicates no substantial curvature, a similarly weak dependence on photon energy cannot be excluded for tumour induction in man.

(38) The increased effectiveness of low-energy x rays has found particular attention with regard to risk–benefit considerations for mammography. The relatively large RBE_M for mammography x rays relative to ^{60}Co γ rays is, therefore, notable. Added interest has been focused on this issue because of a recent claim of an RBE_M in excess of 3 for mammography x rays relative to conventional x rays, based on an experiment on cell transformations in a human hybrid cell line and on re-interpretation of various earlier RBE data (Frankenberg et al., 2002). However, there have been strong objections (Schmid, 2002) against the experimental investigation as well as the survey of RBE data.

(39) The data for dicentrics appear to provide the largest reliable set of RBE_M values for conventional and soft x rays against γ rays. It is, thus, evidently of interest

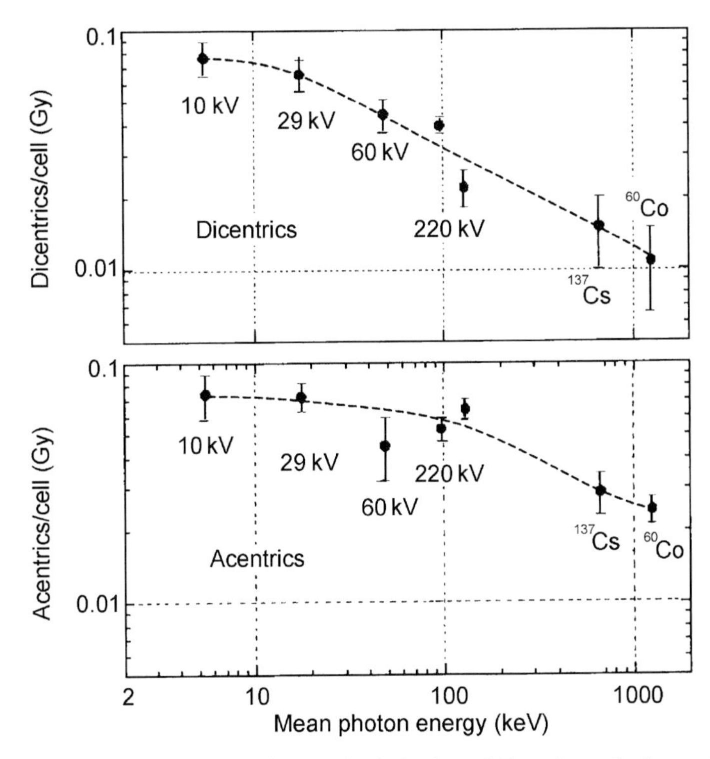

Fig. 2.2. The α coefficients with standard errors for induction of dicentrics and of acentrics in human lymphocytes from the same donor by different x and γ rays in vitro (Schmid et al., 2002b). The broken curves are inserted for visual guidance.

to assess these findings in terms of LET and microdosimetric data, and to compare them with the radio-epidemiological evidence for health effects, such as breast cancer.

2.2.2. Biophysical considerations

(40) Radiobiological determinations of RBE have frequently been related to LET. For protons and heavy ions, this is fairly straightforward because cell studies can be performed with mono-energetic particles in so-called track-segment experiments with comparatively well-defined LET values. In experiments with x rays or electrons, the situation is different because there is always a broad spectrum of LET values (ICRU, 1970) and reference is usually made to the dose-average LET or to the related microdosimetric parameter, dose-averaged lineal energy, y_D.

(41) Figure 2.3 gives computed values for the dose-average LET for the electrons released by mono-energetic photons (solid curves) and also the values for mammography x rays and 200 kV x rays (solid circles and squares, respectively) (Kellerer, 2002). In addition to the dose average, L_D, of the unrestricted LET, the diagram contains the dose averages, $L_{D,\Delta}$, of the restricted LET, L_Δ. L_Δ treats δ rays beyond the specified cut-off energy, Δ, as separate tracks. This accounts in an approximate way for the increased local energies due to δ rays, and therefore provides larger values that are more meaningful than those of unrestricted LET.

(42) The local maximum LET at about 60 keV reflects an important aspect with regard to the effectiveness of x rays. High-energy photons, e.g. ^{60}Co γ rays, release Compton electrons of comparatively high energy and correspondingly low LET. Photons of less energy, e.g. conventional 200 kV x rays, produce less energetic Compton electrons with higher LET. However, as the photon energy is further reduced, the photoelectric effect—i.e. the total transfer of photon energy to electrons—begins

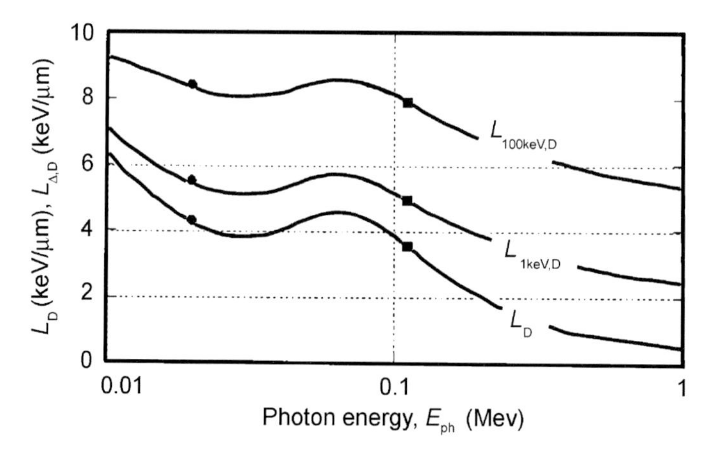

Fig. 2.3. The dose mean restricted and unrestricted linear energy transfer for electrons liberated by mono-energetic photons of energy E_{ph}. The circles and squares give values for 30 and 200 kVp x rays, respectively. They are plotted at the weighted photon energies of the x-ray spectra (Kellerer, 2002).

to dominate, and accordingly the average energy of the electrons begins to increase again. This explains the local minimum of the LET at a photon energy of about 30 keV. Below this energy, the LET increases anew as the photo-electrons have less and less energy. This complexity accounts for the notable fact that, overall, LET does not differ greatly for x rays between 200 and 30 keV.

(43) $L_{D,\Delta}$ is a parameter that correlates with the low dose effectiveness of photon or electron radiation. With a cut-off value $\Delta = 1$ keV, the numerical values of $L_{D,\Delta}$ are consistent with a low-dose RBE of about 2 of conventional x rays vs γ rays. A similar dependence on photon energy is seen in the related microdosimetric parameter dose, lineal energy, y, which has been used as a reference parameter by the ICRP–ICRU Liaison Committee on 'The Quality Factor in Radiation Protection' (ICRU, 1983). Figure 2.4 gives values of y_D as measured by Dvorak and Kliauga (1976) for various photon radiations and for different simulated site diameters, *d*.

(44) The γ rays from the atomic bomb explosions had a mean energy in the range of 2–5 MeV at the distances relevant for survivors (Straume, 1996). This energy range is not covered in Figs 2.3 and 2.4. However, it is apparent from Fig. 2.3 that the mean L_Δ values do not decrease substantially beyond the photon energy 1 MeV. In the microdosimetric parameters, too, there is little change with photon energy above 1 MeV; Lindborg (1976) obtained the value $y_D = 2.1$ keV/μm for 42 MeV photons, which is very close to his value $y_D = 2.4$ keV/μm for ^{60}Co γ rays (see also ICRU, 1983). There is, thus, little evidence that the hard γ rays from the atomic bombs should have an RBE substantially less than 1 compared with ^{60}Co γ rays.

(45) The RBE_M value of about 6 in Fig. 2.2 for 29 kV x rays vs ^{60}Co γ rays indicates—in terms of Fig. 2.3—a correlation with $L_{D,\Delta}$ for a cut-off value in excess of 1 keV. But the essential conclusion from Fig. 2.3 is that the LET values are reasonably consistent with the low-dose RBE values obtained with various photon radiations for chromosome aberrations.

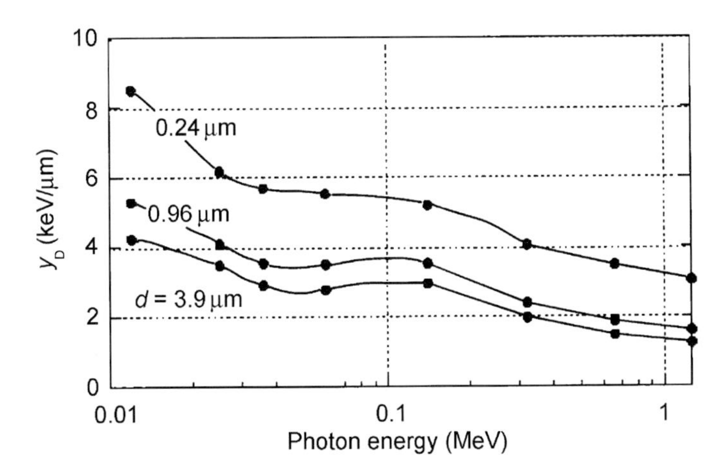

Fig. 2.4. Measured dose-average lineal energy, y_D, for mono-energetic photons and for different simulated site diameters, *d* (Dvorak and Kliauga, 1978).

(46) Figures 2.3 and 2.4 confirm that it would be difficult to explain the recently claimed RBE_M larger than 3 for mammography x rays against conventional x rays. Instead, they suggest a value between 1 and 2, and this is in agreement with an analysis by Brenner et al. (2002) in terms of the microdosimetric data which is supported by new transformation data suggesting an RBE_M of about 1.3.

(47) The analysis (Kellerer, 2002) in terms of the explicit electron spectra at different photon energies leads to the conclusion that the RBE_M of mammography x rays compared with conventional x rays will, regardless of the underlying mechanisms, not exceed 2; this includes a consideration of the potential effect contribution of the 0.5 keV Auger electrons from oxygen that accompany all photo-electrons in water, but only a minority of the Compton electrons which predominate at the higher photon energies.

2.2.3. Information from radio-epidemiology

(48) The follow-up of the atomic bomb survivors has become the major basis of risk estimates for γ rays. Numerous epidemiological studies on medical cohorts have provided risk estimates that exhibit considerable variations. Many of these studies on patients relate to x-ray exposures, but there is no consistent epidemiological evidence for larger risk factors from x rays than γ rays. In fact, while the risk estimates from medical studies are not inconsistent with those for the atomic bomb survivors, they tend to be, as a whole, somewhat lower (UNSCEAR, 2000). The radiation-related increase of breast cancer incidence can serve as an example because it has been studied most thoroughly, and also because it is central to the recent discussions on mammography screening.

(49) Figure 2.5 gives the risk estimates obtained in the major studies on radiation-induced breast cancer. The estimated risk coefficients (and 90% confidence ranges) are expressed in terms of the excess relative risk (ERR) per Gy and the excess absolute risk (EAR) per Gy and 10,000 person years.

(50) The uncertainties are large and the risk estimates vary widely. This is not surprising given that in studies on patients, the patient treatment regimes differ not just in terms of the type of radiation but also in terms of the various exposure modalities, such as acute, fractionated, or protracted exposure, whole or partial body exposure, exposure rate, and magnitude of exposure. Equally important are ethnic differences, including lifestyle-related differences, that are associated with greatly different background breast cancer rates. Thus, there is, even now, a factor of about 6 between the low rates in Japan and the high rates in most Western populations. Populations with low spontaneous rates tend to exhibit comparatively large ERR while their EAR tends to be low. This complicates the comparison of risk estimates, since it remains uncertain whether relative or absolute excess incidence is the more appropriate measure of risk.

(51) It also needs to be noted that the various exposed cohorts differ greatly with regard to the duration of follow-up and, especially, the age at exposure. The last two studies in Fig. 2.5 relate to exposures in childhood, while the other studies refer to exposures at intermediate or higher ages. The last factor is especially critical,

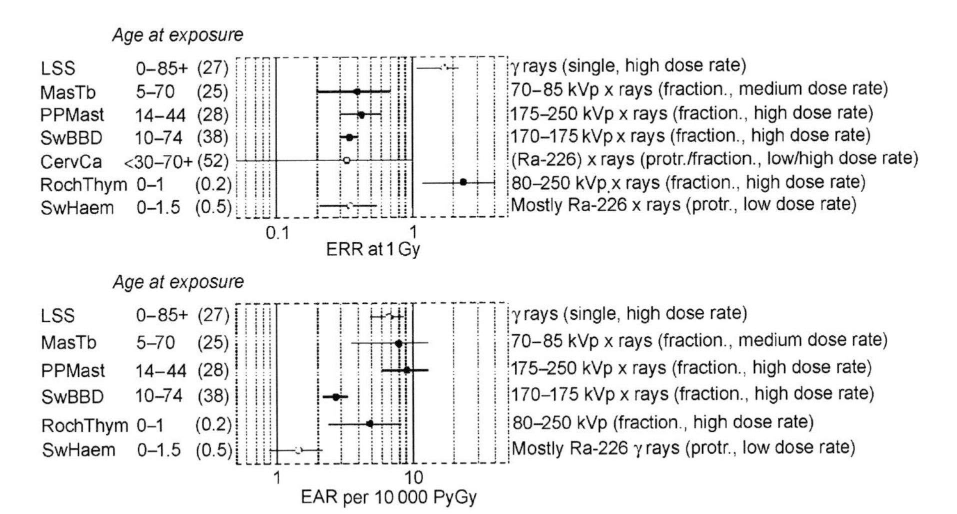

Fig. 2.5. The excess relative risk (and 90% confidence range) from various epidemiological studies on breast cancer. The upper panel gives the excess relative risk per Gy, and the lower panel gives the absolute risk per 10,000 person years and Gy [for description of the individual studies, see UNSCEAR (2000) and Preston et al. (2002)]. The confidence limit for the study on cervix carcinoma patients is recalculated. Cohorts: LSS, Life span study of atomic bomb survivors; MasTb, Massachusetts tuberculosis patients; PPMast, New York post-partum mastitis patients; SwBBD, Swedish benign breast disease patients; CervCa, cervical cancer patients (case control study); RochThym, Rochester infants with thymic enlargements; SwHaem, Swedish infants with skin haemangioma.

because both ERR and lifetime-integrated EAR decrease very substantially with increasing age at exposure.

(52) The dominant influence of the various modifying factors makes it impossible to confirm the difference in effectiveness between γ rays and x rays, or the difference between x rays of different energies on the basis of the epidemiological data. Studies related to other types of cancer are even further removed from providing an answer.

(53) The conclusion is, thus, that cell studies and biophysical considerations suggest an RBE of conventional x rays against hard γ rays of about 2–3, but that this difference has not been confirmed through epidemiological investigations. Therefore, the recommendation of the ICRP to attribute the same w_R (i.e. 1) for γ rays, x rays, and electrons remains a matter of practicability in the absence of definitive information.

2.2.4. The special case of Auger electron emitters

(54) Auger electrons emitted from nuclei bound to DNA are an exception from the *Publication 60* (ICRP, 1991) recommendation for electrons:

> 'Auger electrons emitted from nuclei bound to DNA present a special problem because it is not realistic to average the absorbed dose over the whole mass of DNA as would be required by the present definition of equivalent dose. The effects of Auger electrons have to be assessed by the techniques of microdosimetry.'

The reference to averaging the 'absorbed dose over the whole mass of DNA' may not be entirely clear, but otherwise the argument is convincing. For Auger emitters incorporated into DNA, RBE values between 20 and 40 have been found in transformation studies (Chan and Little, 1986), and calculations of energy-deposition patterns have made the high values plausible (Baverstock and Charlton, 1988; Charlton, 1988). However, at present, there is no recommendation regarding how the microdosimetric assessment is to be performed. y_D can be a suitable substitute for LET (ICRU, 1986), but there is no convention on the size of the reference region in which y is to be measured. Therefore, the issue requires further study.

(55) For Auger emitters bound to DNA, high RBE values have been reported and a w_R of 20 or more appears to be appropriate. For those Auger electron emitters that enter the cell but are not bound to DNA, RBE values between 1.5 and 8 have been found for different endpoints in cell studies (Kassis et al., 1987; Makrigiorgios et al., 1990). Even the less critical case of more uniformly distributed Auger emitters should, therefore, be included in future considerations on a convention for w_R values for Auger electron emitters.

2.3. Different uses of the concept of RBE

2.3.1. RBE as a low-dose equivalence factor

(56) The concept of RBE is treated in this report because it guides the selection of weighting factors for radiation quality. The weighting factors are required to define effective dose, E, which serves as a common scale in radiation protection by attributing equal values to different exposure situations that are presumed to carry roughly the same risk with regard to stochastic effects. Accordingly, the RBE serves in this report strictly as an equivalence factor which is equally dependent on the two types of radiation that are being compared.

(57) The weighting factors relate to low doses, so the present report is—with regard to stochastic radiation effects, such as cancer—primarily concerned with the limit values which the RBE reaches at low doses. These limit values are denoted by RBE_M.

(58) With reference to low doses, it is particularly important to recognise the dependence of RBE on both types of radiation that are being compared. While a high neutron RBE at low doses tends to be seen as an indication of high absolute effectiveness of the neutrons per unit dose, the increase of neutron RBE at low doses is in fact primarily due to the reduced effectiveness per unit dose of γ rays. At low doses, low-LET radiation exhibits a much stronger dependence on modifying factors, such as dose rate or fractionation, than high-LET radiation.

(59) There are practical problems in the determination of RBE at low doses. The major difficulty is the determination of the initial slope of the dose–response curve for single doses of low-LET radiation. Assuming that the response to the reference radiation can be fitted to the linear-quadratic model, the use of low-dose-rate exposures to both the radiation under study and the reference radiation can obviate this problem. Different approaches have been proposed to overcome this problem. They are discussed in Section 3.1.

2.3.2. Dependence of RBE on dose

(60) All experimental and epidemiological information on RBE derives from observations at doses that are considerably larger than the doses of interest in routine radiation protection practice. Assessments of RBE_M must, therefore, combine observed values of RBE with inferences on the increase of RBE toward its limit, RBE_M, at low dose.

(61) Different dependencies of RBE on dose have been found in different systems, but there appears to be a fairly general pattern that can, at low to intermediate doses, be related to a linear dose dependence for densely ionising radiation, such as neutrons, and a linear-quadratic relationship for the low-LET reference radiation, such as γ rays:

$$E_n(D_n) = \alpha_n D_n \quad \text{and} \qquad E_\gamma(D_\gamma) = \alpha_\gamma D_\gamma + \beta_\gamma D_\gamma^2 \tag{2.1}$$

where E_n is the effect due to the high-LET exposure with absorbed dose D_n and E_γ is the effect from the low-LET exposure with absorbed dose D_γ.

(62) The linear and linear-quadratic dose relationships for high- and low-LET radiations were first invoked and explained with regard to exchange chromosome aberrations (Sax, 1938: Lea, 1946). A later analysis of various radiobiological investigations with neutrons in terms of microdosimetry (Kellerer and Rossi, 1972) related the linear and linear-quadratic dependencies more generally to the primary lesions that determine the observed effect, including those cases where the observed effect need not be proportional to the underlying lesions and may even have to be expressed, as is the case with lens opacification (Bateman et al., 1972; Di Paola et al., 1980; Worgul et al., 1996), on an arbitrary scale. The essential point is that the dose–effect relationships for the observed endpoint, e.g. for late effects in man, can involve complexities at the cellular or tissue level which cancel in the comparison of two types of radiation. The RBE is, under this condition, still governed by the linear and linear-quadratic dose dependence of the primary lesions. The Joint ICRP–ICRU Task Group on 'The Quality Factor in Radiation Protection' (ICRU, 1986) emphasised the resulting dose dependence of RBE and the need to consider the low dose extrapolation of RBE in order to estimate RBE_M.

(63) Let D_n and D_γ, respectively, be the high- and low-LET absorbed doses that produce equal levels of effect. The RBE against the dose D_γ of the low-LET reference radiation is then:

$$R(D_\gamma) = D_\gamma / D_n \tag{2.2}$$

and Eq(2.1) provides the relationship:

$$\alpha_n D_n = \alpha_\gamma D_\gamma + \beta_\gamma D_\gamma^2 \tag{2.3}$$

With the substitution $RBE_M = \alpha_n/\alpha_\gamma$, one obtains RBE either as a function of the γ-ray dose or as a, somewhat more complicated, function of the neutron dose:[1]

[1] In the earlier literature, the expression for $R(D_n)$ is of different analytical form because a quadratic term had been included in the dose dependence for the high-LET radiation. Such a term has not been established in experiments, and it is irrelevant except at very high neutron doses.

$$R(D_\gamma) = \mathrm{RBE_M}/(1 + D_\gamma/(\alpha_\gamma/\beta_\gamma)) \tag{2.4}$$

$$R(D_n) = (\surd(1 + 4D_n\mathrm{RBE_M}/(\alpha_\gamma/\beta_\gamma)) - 1)/2D_n(\alpha_\gamma/\beta_\gamma) \tag{2.5}$$

(64) It has been usual to specify or plot the RBE of neutrons as a function of the neutron dose (Kellerer and Rossi, 1972, 1982; ICRU, 1986; Rossi and Zaider, 1996). However, in applications to radiation protection, the quantity of interest is the weighted dose, which is numerically equal to the photon dose. It is, therefore, more informative to specify the RBE as a function of the photon dose. This convention is employed throughout this document.

(65) Figure 2.6 exemplifies the dependency of RBE on the photon and neutron doses for dicentric chromosomes in human lymphocytes and for mutations in *Tradescantia*. The upper curve in the two diagrams refers to the same chromosome data

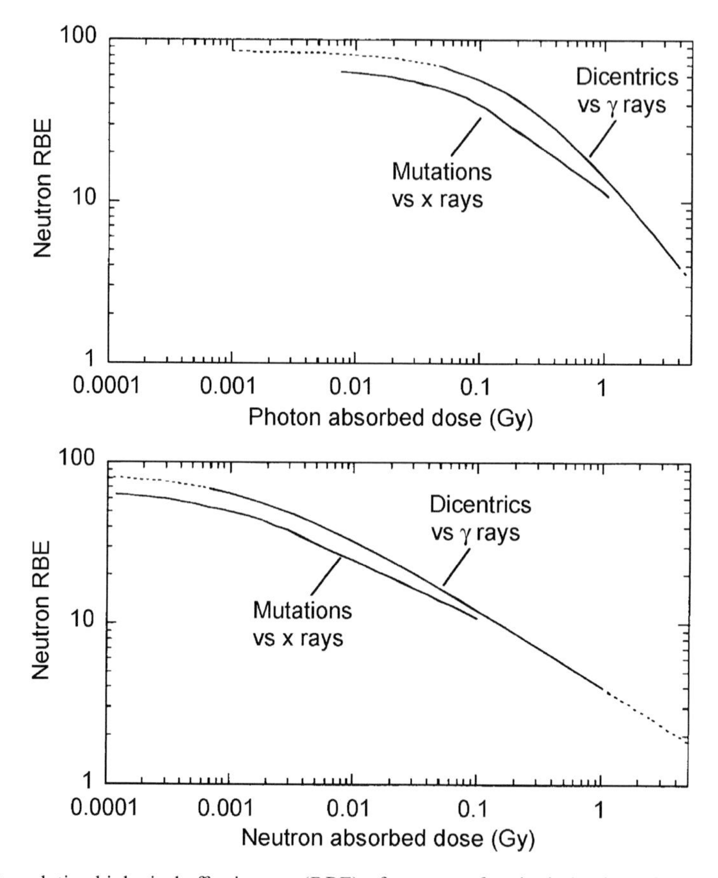

Fig. 2.6. The relative biological effectiveness (RBE) of neutrons for the induction of dicentric chromosomes (upper curves) (Schmid et al., 1998, 2000, 2002a, 2003) and for mutations in *Tradescantia* (lower curves) (Sparrow et al., 1972; see also Fig. 3.4) as a function either of the photon- (upper panel) or the neutron-absorbed dose (lower panel). Dotted sections of the curves are extrapolations beyond the actual data. The chromosome data refer to ^{60}Co γ rays and 0.39 MeV neutrons with the same parameters as in Fig. 3.3; the mutation data refer to 0.43 MeV neutrons and 250 kV x rays.

by Schmid et al. that are represented in Figs 2.2 and 3.3. The lower curve refers to the data by Sparrow et al. (1972) that are represented in Fig. 3.4.

(66) The study on *Tradescantia* is exceptional insofar as it actually shows the plateau of the RBE values at very low doses, and thus provides the RBE_M, which in this case is about 60. In the chromosome studies, there is strong but less direct evidence for an RBE_M value of about 80 for the 0.39 MeV neutrons against the ^{60}Co γ rays.

(67) In some animal experiments with single acute doses, it has not been possible to assess an excess tumour incidence below 1 Gy of the low-LET reference radiation. It is, therefore, difficult in such experiments to ascertain the low dose limit of the RBE. To estimate, nevertheless, the RBE_M from the observed RBE at higher dose, $R(D_\gamma)$, an assumed value of the 'crossover dose', $\alpha_\gamma/\beta_\gamma$, needs to be used in the relationship [see Eq(2.4)]:

$$RBE_M = \left(1 + D_\gamma/(\alpha_\gamma/\beta_\gamma)\right) \cdot R(D_\gamma) \tag{2.6}$$

(68) As noted earlier (ICRU, 1986; UNSCEAR, 1993), the term $\left(1 + D_\gamma/(\alpha_\gamma/\beta_\gamma)\right)$ corresponds to the notion of the dose and dose-rate effectiveness factor (DDREF). This issue will be taken up in Sections 3.1 and 5.2.

(69) On the other hand, it is seen from Eq(2.4) that the observed RBE will approach RBE_M if low dose rates or the low doses in fractionated exposures reduce the coefficient β_γ, i.e. the quadratic term in the response to the low-LET radiation, sufficiently. This fact is used extensively in the studies of life shortening in mice (Section 3.2).

2.3.3. Derivation of the high-LET risk coefficient

(70) As pointed out in the preceding section, RBE values at higher dose and effect levels are employed in order to infer by extrapolation the low dose limit RBE_M. Apart from this indirect use of RBE at higher dose levels, there are also direct applications to risk quantification. While these other applications of RBE are not central to the subject of this report, they are nevertheless outlined here and in the subsequent section.

(71) Risk estimates for radiation protection purposes are obtained primarily from epidemiological observations on groups of people exposed to substantial doses of x rays or γ rays. The nominal risk coefficient for photon radiation is provided by an extrapolation to low doses or low dose rates that accounts, in terms of the DDREF, for an assumed degree of curvature in the dose relationship and for the related dose-rate dependence (ICRP, 1991). However, it needs to be noted that the risk estimates by extrapolation from higher doses are largely confirmed by the low dose data on cancer mortality and incidence in the A-bomb survivors (Pierce and Preston, 2000).

(72) The determination of risk coefficients on the basis of epidemiological data has also been possible for certain high-LET radiations, such as radon and its decay products (NAS, 1999), radium (see Machinami et al, 1999; Nekolla et al., 2000), or, more recently, plutonium (Kreisheimer et al., 2000; Koshurnikova et al., 2002). However, for other densely ionising radiations, such as fast neutrons, epidemiological information is lacking, and indirect procedures to determine the risk estimate,

C_n, for the high-LET radiation from experimental observations are employed. The common method is to derive, from animal experiments or other radiobiological studies, the RBE_M value for the high-LET radiation and then to multiply this value into the nominal risk coefficient, C_γ, for photons:[2]

$$C_n = RBE_M \cdot C_\gamma \tag{2.7}$$

(73) The two quantities RBE_M and C_γ need to be determined from low-dose or low-dose-rate data. With this low-dose procedure, the risk estimate $C_n = RBE_M \cdot C_\gamma$ is, thus, the product of two numerical values that are both based on extrapolation. The estimate, C_n, obtained by the low-dose procedure is, accordingly, subject to considerable uncertainty.

(74) A more direct high-dose method has been employed to derive the risk estimate for fission neutrons (Kellerer and Walsh, 2001). The approach requires the excess relative risk, ERR_{ref}, observed among the atomic bomb survivors at a high reference γ-ray dose, $D_{\gamma,ref}$. The observed effect-to-dose ratio is termed the reference slope (slope of the dotted line in Fig. 2.7):

$$c_{\gamma,ref} = ERR_{ref}/D_{\gamma,ref} \tag{2.8}$$

(75) The reference slope can be determined with considerably less uncertainty than the putative initial slope, C_γ, of the dose dependence for γ rays (i.e. the solid curve in Fig. 2.7). If an intermediate value of the reference dose is chosen, the reference slope is roughly equal to the slope of the linear fit to the γ-ray data.

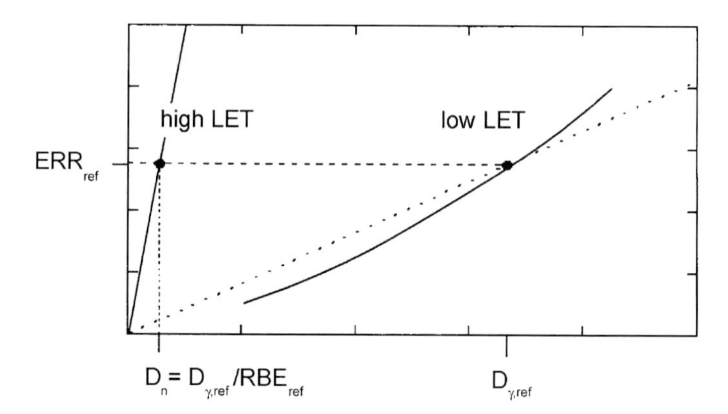

Fig. 2.7. Diagram of excess relative risk vs dose to explain the derivation of the high-linear-energy-transfer (LET) risk coefficient without recourse to the low-dose relative biological effectiveness (RBE). The low-dose part of the low-LET dose–effect relationship is omitted in order to emphasise the point that it is not reliably known and that it is not required for the specified approach.

[2] The notation C_γ rather than α_γ is used here for the initial slope of the dose dependence of risk (e.g. in terms of ERR) to make the distinction between epidemiological results and experimental radiobiological data.

(76) Furthermore, the method requires the RBE_{ref} of fission neutrons as observed in rodents against the same γ-ray reference dose. If the RBE for rodents also applies to man, the neutron dose $D_{\gamma,ref}/RBE_{ref}$ will cause the same ERR_{ref} as the γ-ray dose $D_{\gamma,ref}$. With the usual assumption of linearity for neutrons, the slope of the neutron dose-effect curve (straight solid line in Fig. 2.7), i.e. the risk estimate for the neutrons, is then:[3]

$$c_n = ERR_{ref}/\left(D_{\gamma,ref}/RBE_{ref}\right) = RBE_{ref} \cdot c_{\gamma,ref} \qquad (2.9)$$

This estimate c_n is, unlike the familiar low-dose estimate, C_n, based on two observed, rather than two extrapolated, quantities.

(77) A suitable reference dose with regard to the data from the atomic bomb survivors happens to be $D_{\gamma,ref} = 1$ Gy (Kellerer and Walsh, 2001, 2002). Accordingly, in this high-dose procedure, one needs to refer to an estimate of the RBE from animal studies against the same γ-ray dose of 1 Gy, which tends to be in the lower range of γ-ray dose level for which neutron RBE values have been obtained (see Section 3.3).

(78) Both estimates, C_n and c_n, make sense, but the high-dose formulation in Eq(2.9) has the advantage that there is considerably less statistical uncertainty in the reference slope, c_γ, than in the estimated initial slope, C_γ. Furthermore, there is less uncertainty in RBE_{ref} than in RBE_M. The uncertainty of the neutron risk estimate c_n is, thus, considerably less than the uncertainty of the estimate C_n.

(79) In both the low- and the high-dose method, there is, of course, unavoidable uncertainty in the extrapolation of RBE from experiments in rodents to man, but this needs to be accepted in the absence of direct epidemiological information. There is also the assumption of a linear dose relationship for the high-LET radiation, but up to the neutron doses below 0.1 Gy that are involved in this analysis, it is well supported by radiobiological evidence.

(80) These considerations make it clear that the derivation of weighting factors for radiation quality and the quantification of the risk of high-LET radiation are two distinct issues. The uncertainty of the low dose effectiveness of low-LET radiation is inherent in RBE_M and w_R. It does not enter the high-LET risk estimates if derived in terms of the modified procedure.

(81) In Section 3.1, a low- and a high-dose approach to the derivation of RBE_M will be considered. This approach uses, in addition to an assumed DDREF, a high-dose RBE, RBE_H, which is roughly equivalent to RBE_{ref}.

2.3.4. Derivation of risk under specified conditions

(82) The nominal risk coefficient serves as a general guideline in the setting of dose limits, but it is not meant to be applied in the derivation of quantitative risk estimates under specific conditions. An example is the use of ionising radiation in medical diagnostics. One reason for the inapplicability of the nominal risk coefficient under these conditions is the fact that the age distributions of the exposed groups of people can differ substantially from the age distribution of the general population or

[3] The accounting for the neutron component in the A-bomb radiation is disregarded in this summary explanation of the method.

of a working population, which can alter the risk substantially. Furthermore, in order to obtain a specific risk estimate, one may need to take specific characteristics of the exposed population, such as ethnic factors, into account Finally, in the case of medical applications of ionising radiation, exposures are commonly organ specific. It is then necessary to base numerical evaluations of risk on organ doses, organ-specific risk estimates, and, where required, on specific estimates of the RBE.

(83) In view of these considerations, there is no conflict between the fact that all photon radiations are, with the current convention for effective dose, given the same weight, while risks from medical applications of soft x rays, conventional x rays, and hard γ rays are assessed differently. **w_R is designed for the practice of radiological protection, not for specific risk assessment**. Even the RBE values from experimental systems have limited applicability to risk assessment. For example, it would be inappropriate to base cost–benefit considerations for mammography screening primarily on risk estimates for γ rays and RBE values, rather than using the more relevant epidemiological data for x rays.

(84) The determination of probabilities of causation (PC) is another example of risk assessment under specified exposure conditions. (NIH, 1985; Kocher, 2001; Kocher et al., 2002; Land et al., 2002). PC values are computed for intermediate- or high-dose exposures that cause risks of a magnitude comparable to the background risk. This means that neither the values of RBE_M nor the weighting factor conventions w_R and $Q(L)$ are applicable; instead, high-dose information needs to be applied.

(85) Kocher et al. (2002) presented a synopsis of proposed w_R values for use in calculating probabilities of causation of cancers. Since this particular use of RBE data differs from the use for regulatory purposes, the authors state:

> 'We have decided to abandon the term 'relative biological effectiveness (RBE) factor' to describe the biological effectiveness of different radiation types for the purpose of estimating cancer risks and probability of causation, basically to avoid misuse of the term 'RBE', which should be used only to describe the results of specific radiobiological studies. We also should not use the term 'radiation weighting factor' because this is an ICRP-defined point quantity used in radiation protection (i.e. to calculate equivalent doses). Therefore, we need a new term that is short and reasonably descriptive. We propose to describe the quantity of interest by the term 'radiation effectiveness factor (REF).'

(86) A clear distinction of the various types of RBE and the quantities derived from RBE is essential. The term 'radiation effectiveness factor (REF)' was employed earlier (ICRP, 1984) with a different connotation to designate the ratio of the high-dose value of the RBE for deterministic effects to its maximum value, RBE_m, at low doses. In this earlier definition, REF was the analogue for deterministic effects to the DDREF for stochastic effects (see Section 5.2.2). However, the earlier definition was no longer used in *Publication 58* (ICRP, 1990). Accordingly, the proposal of Kocher et al. (2002) appears suitable, and it is recommended that the term REF be used for numbers that replace, by convention, RBE values in computations of the probability of causation.

3. QUANTIFICATION OF RBE

3.1. Two approaches to the determination of RBE_M

(87) *Publication 60* (ICRP, 1991, paragraph 32) stated:

'Both equivalent dose and effective dose are quantities for use in radiological protection, including the assessment of risks in general terms. They provide a basis for estimating the probability of stochastic effects only for absorbed doses well below the thresholds for deterministic effects. For the estimation of the likely consequences of an exposure of a known population, it will sometimes be better to use absorbed dose and specific data relating to the relative biological effectiveness of the radiations concerned and the probability coefficients relating to the exposed population.'

Earlier in *Publication 60,* the ICRP (1991) stated that the radiation weighting factor (w_R) value for a specific type and energy of radiation was selected by the Commission to be representative of values of the relative biological effectiveness (RBE) of that radiation in inducing stochastic effects 'at low doses', and it went on to say 'The RBE of one radiation compared with another is the inverse ratio of the absorbed doses producing the same effect'.

3.1.1. The low-dose method

(88) It was understood that for the purpose of defining an effective dose, E, and the organ-equivalent doses, H_T, values of the quality factor $Q(L)$ and w_R values were required that had to be related to the RBE of high-linear-energy-transfer (LET) radiations—such as neutrons—for low doses and/or low dose rates. Since the RBE is largest at low dose and low dose rate, the relevant values were termed RBE_M.

(89) The low-dose method used by the National Council on Radiation Protection (NCRP, 1990) was to determine RBE_M as the ratio of the initial slopes of the dose–response curves of the induction of tumours by fission neutrons and the reference radiation, γ rays.[4] Values were reported for various types of tumour in two strains of mouse. Due to the difficulty in determining the initial linear slopes of the dose responses, especially of the reference radiation, the errors are large.

[4] The value of the ratio was denoted RBE_M in the ICRU–ICRP report (1963) and NCRP Report 64 (NCRP, 1980). In 1980, the NCRP reviewed the experimental data from which a dose-rate effectiveness factor (DREF) could be derived:

DREF = effect per unit dose at high dose and dose rate/effect per unit dose at a low dose rate where the effect per unit dose represents the linear regression coefficients obtained for the high-dose-rate and low-dose-rate data for cancer induction. ICRP (1991) introduced the term 'dose and dose-rate effectiveness factor (DDREF)' which implied that the dose–response relationship was linear quadratic.

3.1.2. The high-dose method

(90) To remove the need to extrapolate the RBE in every single experimental system, the Committee on Interagency Radiation Research and Policy Co-ordination (CIRRPC) suggested a high-dose method which uses the observed high-dose RBE, RBE_H, and extrapolates it to low dose by the standard modifying factor (DDREF) (ICRP, 1991) which is inferred from the entirety of data relevant to late effects in man. As stated by the CIRRPC (1995):

> '... the RBE_M for humans is determined by the product of the appropriate neutron value determined under high-dose/dose-rate conditions and the DDREF selected for use with human low-LET acute exposure risk coefficients...Thus, the high-dose/dose-rate data (RBE_H) is modified for use in the case of low-dose/dose-rate exposure (RBE_M) by the DDREF selected for use with human low-LET acute-exposure risk coefficients, or:
>
> $$RBE_M = RBE_H \cdot DDREF \ldots' \quad (3.1)$$

(91) It was pointed out that, in principle, both methods should result in the same 'value of RBE_M for human beings'. However, it was stated that the high-dose approach has the advantage of being based on high-dose and high-dose-rate data where the uncertainty of laboratory RBE values is smaller and less dependent on the choice of exposure conditions for the reference radiation.

(92) In response to a request by the British Committee on Radiation Units and Measurements (BCRU), the National Radiobiological Protection Board (NRPB) carried out a review of the RBE values and how they might be applied to the problem of deriving equivalent doses.

(93) The NRPB accepted the approach that had been proposed by the CIRRPC (1995) and made it more specific. It equated the RBE_H value of the high-dose RBE, in Eq(3.1), with the ratio of the initial slope (H) of the dose-response curve for the induction of cancer by high-LET radiation, and the slope (LL) of the linear fit to intermediate- and high-dose data for cancer induction by low-LET radiation.[5] It selected—in line with the choice in *Publication 60* (ICRP, 1991) and with the subsequent recommendations by the NRPB (1997)—a value of 2 for the DDREF:

$$RBE_A = DDREF.RBE_H = 2(H/LL) \quad (3.2)$$

(94) The NRPB chose the new notation RBE_A in order to indicate that this RBE_M value is derived through the specific approach represented by Eqs(3.1) and (3.2). As stated, the approach is an attempt to avoid the need to perform low-dose or low-

[5] The unfamiliar notation LS, H, and LL for the slopes of the linear fit to the dose–effect relationship is retained solely for reference to the 'high-dose method' proposed by the CIRRPC panel and the NRPB in this section and in the following subsection.

dose-rate observations in each RBE assessment to determine the initial slope of the dose response of low-LET radiation. Instead, the approach uses an assumed standard value for DDREF.

(95) Values of DDREF based on experimental animal data were considered by the NCRP (1980) and by UNSCEAR (1988, 1993) to range between 2 and 10 with large associated errors. However, there are very few types of cancer for which adequate data have been obtained with low dose rates, and certainly such data are not available for a representative spectrum of relevant tumours for either mice or rats. In view of these limitations, it is not a priori clear which of the two methods to derive RBE_M is more reliable.

(96) In summary, obviously problems arise from not having epidemiological data for the effects of neutrons and having to rely on experimental data. However, despite the inherent uncertainties, the use of experimental data is unavoidable for both neutrons and heavy ions. The issue is considered in more detail in Chapter 4.

3.1.3. Relationship to risk estimation for high-LET radiation

(97) The preceding considerations were related to the derivation of the limiting RBE at low doses. But there is also a relationship between the high-dose method for the derivation of RBE_A and the high-dose method for the derivation of the high-LET risk coefficient (Section 2.3.3). As will be seen, the difference lies in the fact that the factor DDREF is not required for the high-dose risk estimate.

(98) Let LS be the slope of the linear fit to the data from the atomic bomb survivors and c_n the risk estimate for neutrons. Prior to the emergence of direct low-dose risk estimates from the follow-up of the A-bomb survivors (Pierce and Preston, 2000), the CIRRPC panel noted that the coefficient for human risk estimates for low-LET radiation is derived from the high-dose and high-dose-rate data (LS) from the atomic bomb survivors modified by the DDREF. To this corrected risk estimate, RBE_A values are applied to obtain c_n. The procedure is expressed by the relationship [see Eq(3.2)]:

$$c_\mathrm{n} = (LS/\mathrm{DDREF})\mathrm{RBE}_\mathrm{A} = (LS/\mathrm{DDREF}) \cdot \mathrm{DDREF} \cdot (H/LL) \qquad (3.3)$$

(99) If, as suggested by the CIRRPC panel and the NRPB, the same standard modifier DDREF is applied in the derivation of the corrected low-LET risk estimate ($= LS/\mathrm{DDREF}$) and of the value $\mathrm{RBE}_\mathrm{A}[= \mathrm{DDREF}(H/LL)]$, the factor DDREF cancels in Eq(3.3). With $\mathrm{RBE}_\mathrm{H} = H/LL$, one obtains from Eq(3.3):

$$c_\mathrm{n} = \mathrm{RBE}_\mathrm{H} \cdot LS \qquad (3.4)$$

This is equivalent to Eq(2.7) in Section 2.3.3 for the high-LET risk coefficient:

$$c_\mathrm{n} = \mathrm{RBE}_\mathrm{ref} \cdot c_{\gamma,\mathrm{ref}} \qquad (3.5)$$

The difference is that RBE_H and LS are somewhat vaguely related to 'intermediate- and high-dose data', while the analogous parameters RBE_{ref} and $c_{\gamma,ref}$ are related to a specified γ-ray reference dose.

(100) ICRP has emphasised in *Publication 60* (1991, paragraph B62) that DDREF needs to be based on a broad judgement of the entirety of epidemiological and experimental data. It explained that a seeming discrepancy between larger DDREF values from animal experiments and the lower DDREF value suggested by the human data reflects, at least in part, the fact that the reference doses tend to be larger in animal experiments than in epidemiology. The expression $[1 + D_\gamma/(\alpha_\gamma/\beta_\gamma)]$ for DDREF in Eq(2.4) gives numerical guidance. With crossover dose $\alpha_\gamma/\beta_\gamma = 1$ Gy, DDREF = 2 for a reference dose $D_{\gamma,ref} = 1$Gy, and for $D_{\gamma,ref} = 3$ Gy, DDREF = 4 (see also UNSCEAR, 1993). In recommending a value of 2 to be used for the DDREF, ICRP recognised that the choice is somewhat arbitrary and may be conservative.

3.2. The use of life shortening in the determination of RBE_M

(101) The cost and complexity of animal studies on radiation-induced tumours make it attractive to use alternatives that provide, in a simpler way, largely equivalent information. The major possibility is determination of radiation-induced life shortening. For mice, life-shortening data are available that can be considered as a suitable endpoint for obtaining a direct estimate of RBE_M. The advantages and disadvantages of using life shortening are as follows.

3.2.1. Advantages

- At low dose rates, a very high percentage of life shortening is attributed to excess cancers.
- Life shortening is an integral of the effect of all the types of cancer that occur in any one strain of mouse studied. The contribution of non-cancer effects also increases with dose and dose rate, but this is not substantial until either the dose or dose rate has reached a certain level.
- Dose rates at which the effect becomes dose-rate independent can be used.
- Exposures that are protracted over a considerable period but not all of the life span (terminated exposures) can obviate the criticism of 'wasted radiation' levelled at duration-of-life irradiation. Multiple small fractions can be used for both the neutron or γ radiations.
- The slopes of the dose–response curves for life shortening by both neutrons and γ rays are linear, at least up to about 1 Gy total dose of neutrons and up to higher doses with γ rays.
- There are virtually no errors in diagnosis and none of the problems of deciding whether a tumour is lethal or what is the cause of death.
- The analysis is easier than for cancer induction for which independence is assumed and seldom verified. No corrections for competing risks need to be made.

- Life shortening provides data for a single RBE_M. While the spectrum of tumours is somewhat restricted, it obviates the problem of weighting individual RBE_M values.

3.2.2. Disadvantages

- The spectrum of cancers that contribute to life shortening is determined by the susceptibility characteristics of the strain of mouse or rat used.
- With low doses and low dose rates, the cancers appearing later in life tend to predominate.
- Life-shortening studies, as well as most studies on radiogenic cancers, are carried out on a population of experimental animals that are of one age, in contrast to the broad age distribution of a general human population.
- Studies of radiation-induced cancer in mice that provide quantitative data guiding radiation protection decisions should be carried out on male mice.[6] This, of course, means that cancers of the breast and other tumours that show a strong influence of gender are not included.

(102) Since the advantages tend to outweigh the disadvantages, life shortening should carry considerable weight when RBE_M values are considered in the selection of the w_R for fission spectrum neutrons and for other radiation qualities. However, there remains the same problem that exists with all data obtained from small animals, namely, that the LET and energy spectrum in the target tissues is very different in humans compared with mice. This issue will be dealt with in Chapter 4.

3.2.3. Determination of RBE_M for life shortening

(103) For risk comparison between different radiations, it is essential that the best possible estimate of RBE_M be made. In the 1963 report of the RBE Committee (ICRU-ICRP, 1963), as noted above, RBE_M was defined as the ratio of the initial slopes of the dose–response curves for the radiation under study and the reference radiation.

(104) It has been difficult, not only for tumour induction but also for life shortening, to establish the initial slopes of the dose–response curves. Earlier life-shortening data that did not include doses below 0.2 Gy of neutrons were compatible with the conclusion that the neutron RBE varies inversely with the square root of the dose (Kellerer and Rossi, 1972, 1982) in an intermediate dose range. More recent analyses by Storer and Mitchell (1984) and Thomson et al. (1983, 1985) indicated a limiting value of RBE. The evidence that small doses of neutrons are additive (Storer and Fry, 1995) and that the life shortening due to γ rays becomes dose-rate independent at about 0.15–0.20 Gy/day (Sacher, 1976) is consistent with a limiting value of RBE against γ-ray doses of this magnitude.

---

[6] Effects on the ovary of the mouse occur at very low doses, resulting in altered hormonal balance and higher probabilities of tumours dependent on sex hormones. There is also an increased probability of multiple tumours.

(105) Thomson et al. (1983) estimated RBE_M values of 20 and 25 for male and female B6CF$_1$ mice, respectively, comparable with 13.3 for female BALB/c mice reported by Storer and Mitchell (1984).

(106) There are a number of data sets for radiation-induced life shortening that illustrate the influence of gender, irradiation pattern, reference radiation, and the method of analysis. It is clear that the dose–response curves for γ radiation, the most commonly used reference radiation, vary considerably with the pattern of irradiation such as fractionation, continuous exposure, and protracted exposures terminated in midlife (Carnes et al., 1989). The study of Covelli et al. (1989) indicates that the neutron RBE for life shortening is, as is expected, lower when x rays, rather than γ rays, are used as reference radiation.

(107) Although RBE_M values for life shortening appear to differ among strains of mouse, perhaps by a factor of 2, this is considerably less than the range of reported RBE_M values among individual types of cancer. The spectrum of cancers is strain dependent, and therefore some strain-dependent difference in life shortening would be expected.

(108) In Table 3.1, the influence of the pattern of irradiation and gender is demonstrated and a value of about 43 can be considered the maximum value for the RBE for life shortening in the hybrid B6CF$_1$ mouse.

(109) It should be noted that augmentation, i.e. the so-called inverse-dose-rate effect, had no influence on the estimation of RBE_M because augmentation only occurs at higher levels of neutron doses. The estimates of RBE_A in Table 3.2 (NRPB, 1997) are based on the same data as the results in Table 3.1. The RBE_A estimates are about two to three times lower than the values for RBE_M. This reflects the use of DDREF=2 in Table 3.2, instead of the higher value of DDREF=4 for life shortening after exposure for 22 h/day, 5 days/week for 59

Table 3.1. Relative biological effectiveness (RBE_M) for life shortening in B6CF$_1$ mice[a] (based on data from Carnes et al., 1989)

	No. of exposures	$RBE_M \pm SE$
Females		
	1	12±2
	60 (1/week)	43±6
Males		
	1	8±2
	60 (1/week)	24±4
	59 (5/week, 22 h each)	42±7

[a] Fission spectrum neutrons, mean energy 0.85 MeV, and ^{60}Co γ rays. $RBE_M = \alpha_H/\alpha_{LQ}$ where α_H is the initial slope of the neutron dose–response curve, and α_{LQ} is the α component of the linear-quadratic fit to the γ-ray dose-response curve.

Table 3.2. Relative biological effectiveness (RBE_A) for life shortening in B6CF$_1$ mice[a] [based on data from Carnes et al. (1989); analysis by Edwards (NRPB, 1997)]

Gender	$RBE_A \pm SE$
Female	19 ± 2
Male	16 ± 2

[a] Fission spectrum neutrons, mean energy 0.85 MeV, and ^{60}Co γ rays. $RBE_A = 2\alpha_H/\alpha_{LL}$ where α_H is the initial slope of the neutron dose–response curve and α_{LL} is the coefficient of a linear fit of the dose response determined with high-dose-rate γ radiation. The α_H values were derived as average from linear fits with three different neutron dose cut-off values (0.1, 0.21, and 0.4 Gy). The notation RBE_A is used to alert one to the fact that the so-called high-dose method (Section 3.1.2) is used.

weeks which was determined by the Argonne National Laboratory Group and is shown in Table 3.1.

(110) There are a number of studies of life shortening from which estimates of RBE have been made, dating back to the 1957 study of Neary et al. at the GLEEP reactor (~0.7 MeV neutrons). The results of all the studies suggest the following:

- For single exposures in studies of seven mouse strains and *Peromyscus leucopus,* the RBE values against γ rays range from about 10 to 16.
- For fractionated and short-term exposures, the RBE values range from about 11 to 30; for duration of life and long-term protracted exposure, the RBE values range from about 17 to 43.

(111) A small number of studies uses 250 kV x rays as the reference radiation. Covelli et al. (1989) estimated the RBE against x rays to be 12 for BC3F female mice exposed to single doses of 1.5 MeV neutrons. Thus, this study suggests that the RBE is lower against x rays than the reported values against γ rays.

(112) Di Majo et al. (1996) estimated the RBE against x rays in male CBA/CNE mice to be 24–47 for neutrons with an average energy of about 0.4 MeV, but 7–9 in female mice. The gender-dependent difference in the neutron RBE may reflect the significant differences in the susceptibility to certain tumours, for example myeloid leukaemia, which occurred exclusively in irradiated male mice at a younger age than many of the solid cancers. The susceptibility for myeloid leukaemia is strain dependent and is high in CBA mice.

(113) There has been no study focused on the neutron RBE for life shortening in rats. For the study at the French Commissariat de l'Energie Aromique (CEA) of neutron RBE for tumour induction, Wolf et al. (2000) derived an RBE of 35 against a γ-ray dose of 1 Gy for male Sprague-Dawley rats in terms of life shortening; the corresponding value for tumour induction was 50 (see Section 3.3.3).

3.2.4. Radiobiological concerns

(114) As noted in Section 2.1, the need of single maximum RBE values for radiation protection purposes was discussed in the ICRP–ICRU RBE report (ICRU–

ICRP, 1963), and the term 'RBE$_M$' was introduced to denote the ratio of the initial slopes of the dose–effect curves for the radiation under study and the reference radiation.

(115) It is not stated in *Publication 60* (ICRP, 1991) which data were used in the selection of the values of w_R. The report says:

> 'the value of the radiation weighting factor for a specified type and energy of radiation has been selected to be representative of the values of the relative biological effectiveness of that radiation in inducing stochastic effects at low doses',

and later:

> 'the Commission now selects radiation weighting factors, w_R, based on a review of the biological information, a variety of exposure circumstances and inspection of traditional calculations of the ambient dose equivalent'.

(116) However, the close connection of the current w_R values to the assessment of radiobiological data by Sinclair (1985), by NCRP Report 104 (1990), and by the report of the Joint ICRP–ICRU Task Group on 'The Quality Factor in Radiation Protection' (ICRU, 1986) needs to be noted. The recommendations of the Joint Task Group were, as explained in Chapter 4, converted into the current $Q(L)$, and the w_R values were then derived by ICRP (1991) in terms of the ambient dose equivalent, q^*.

(117) In this connection, it must be pointed out that the conclusions of the Joint ICRP–ICRU Task Group were primarily focused on cytogenetic data. While an argument can be made for the use of RBE values obtained from studies of the induction of chromosome aberrations, the induction of tumours, and even life shortening in animals, appears more appropriate. In the case of neutrons, most experimental animal data have been obtained for fission spectrum neutrons.

(118) There have been a number of reviews of RBE for tumour induction, life shortening, and cellular effects induced by fission neutrons (Sinclair, 1982; UNSCEAR, 1982; ICRU, 1986; NCRP, 1990; CIRRPC, 1995; NRPB, 1997; Engels and Wambersie, 1998; IARC, 2000).

(119) The NCRP (1990) stressed that an accurate determination of the initial slopes of the dose responses of both the reference radiation and the radiation under study is central to the determination of RBE$_M$. In general, the determination of the initial slope of the dose–response curve of the low-LET reference radiation is more difficult than for high-LET radiations. The best approach, consistent with the assumption of a linear-quadratic fit, is to determine the slope of the response after low-dose-rate exposure or multiple small fractions. It is difficult to quantify the linear coefficient of a linear-quadratic response. For example, despite the inclusion of data points below 0.5 Gy, Ullrich and Preston (1987) could not dismiss a linear fit to the data for the induction of myeloid leukaemia by γ rays or neutrons. The estimate of the RBE for fission neutrons was 2.8 compared with a value of about 16 obtained

with protracted low-dose-rate γ radiation (Upton et al., 1970). The most appropriate experimental approach to the estimation of RBE_M is, thus, that exposure to both the reference radiation and the radiation under study be at low dose rate.

3.3. Neutrons

3.3.1. Mice

(120) The selection of data from studies on mice, from which w_R values are inferred for specific types and energies of ionising radiation, involves a number of aspects, for example:

- in the determination of RBE_M, greater weight should be given to data obtained with low dose rates (at least for the low-LET reference radiation);
- only data for relevant organs and tissues should be considered. For example, the ovary should be excluded since cell killing is an important component of the mechanism of ovarian tumorigenesis;
- data for lethal and non-lethal tumours in a specific organ or tissue should not be pooled, and data for lethal tumours should be adjusted for competing risks.

(121) Unfortunately, there are—among the results that might be used in the selection of w_R values—very few data sets from experiments in mice that meet these criteria for estimating the RBE_M for solid cancer induction. In a critical review of the available data from mice and rats, the NCRP (1990) concluded that the estimates of the RBE_M based on the ratio of the initial slopes of the responses to fission neutrons and γ rays were limited to six tissues in female mice. Ullrich et al. (1976, 1977) reported RBE_M values for thymic lymphomas and for tumours of the pituitary, Harderian gland, and lung tumours in RFM mice, for lung and mammary gland in BALB/c mice (1983, 1984), and myeloid leukaemia in RFM mice (Ullrich and Preston, 1987). The values of RBE_M ranged from about 3–59 with considerable errors. As noted, the murine ovary is exquisitely radiosensitive. Oocytes can be inactivated by relatively low doses, which results in altered hormone balance and changed rates of particular tumours at multiple sites; RBE values determined in female mice must, therefore, be used with caution. Only two of the six types of tumour for which data are reported are germane to risk estimates in humans. RBE_M values for lung and mammary adenocarcinomas in BALB/c mice were 19 ± 6 and 33 ± 12, respectively. The NRPB (1997) re-analysed data for various experiments and estimated comparable RBE_M values (and standard error ranges) from the studies by Ullrich (1984): 20 (12–30) for lung, and 27 (13–41) for mammary tumours. The estimates of RBE_A were 15 (11–20) for lung, and 7 (5–10) and 23 (15–40) for mammary gland in BALB/c mice.

(122) The NRPB estimated RBE_M and RBE_A for vascular tissue and all epithelial tissue except ovary based on the data and analysis reported for $B6CF_1$ mice by Grahn et al. (1992). In this study, linear and linear-quadratic dose–response equations were fitted to the mortality and incidence data for eight different tumours or

groups of tumours. The equations were constrained through the control intercepts and fitted to the data for tumour occurrence and death at 600–799 days and 800–999 days after exposure to Janus reactor neutrons and ^{60}Co γ rays.

(123) Grahn et al. (1992) estimated RBE (calculated as the ratio of linear coefficients) in male mice to be 25$\pm$4 for lung tumours, 26$\pm$4 for all epithelial tumours, and 15$\pm$3 for vascular tumours based on the data for the exposure protracted over about 60 weeks in the 600–799-day group. The NRPB derived estimates of RBE_A and RBE_M from the data of Grahn et al. (1992), and reported RBE_M and RBE_A for vascular tumours of 13.9$\pm$2.6 and 9.4$\pm$3, respectively, and 22.7$\pm$4.7 and 11.0$\pm$1.9, respectively, for all epithelial tumours. The Grahn et al. (1992) values for RBE were obtained from data for lethal and non-lethal tumours detected when the mice were 700–1000 days of age. With low-dose experiments in this long-lived hybrid mouse (maximum life span about 1480 days), the majority of tumours occur late in life. The pooling of lethal and non-lethal tumours, while undesirable, increased the robustness of the statistics. However, there is evidence to suggest that RBE may differ between lethal and non-lethal tumours in mice. The studies of life shortening indicated that no augmentation occurred with protraction below 0.4 Gy, but the linear terms for tumour response to neutrons showed a general upwards trend as the dose was protracted. However, this effect is small compared with the decrease with protraction of the γ rays. RBE for the pooled data for solid cancers, for example, the epithelial tumours, and the three types of tumour analysed individually (lung, liver, and Harderian gland) varied widely from 7 to 100 ($\pm$40) in the 800–999-day group.

(124) Di Majo et al. (1990) determined the age-related susceptibility of $BC3F_1$ mice to the induction of liver tumours by neutrons and 250 kV x rays. The data for age 3 months at exposure are suitable for comparison of the neutron- and x-ray-induced excess incidence; they are represented in the lower panel of Fig. 3.1. There is no recognisable initial slope in the dose dependence for the x rays. Accordingly, the authors fitted their data to a linear dependence for the neutron experiments and a quadratic dependence for the x rays. This is, of course, no proof that RBE_M is infinite in this case. An RBE of about 15 is estimated against an x-ray dose of 2 Gy. The statistical uncertainty of the data for lower doses does not permit more than the conclusion that RBE_M is likely to exceed 15.

(125) The life-shortening data from the same experiment are represented in the upper panel of Fig. 3.1. Here, RBE is about 10 against an x-ray dose of 2 Gy, and the statistical uncertainty precludes a specification beyond the observation that RBE_M appears to exceed 10.

3.3.2. Leukaemia and lymphoma

(126) In mice, RBE values have been determined for thymic lymphoma of T-cell origin (Upton et al., 1970; Ullrich et al., 1976), myeloid leukaemia (Upton et al., 1970; Ullrich and Preston, 1987; Di Majo et al., 1996), malignant lymphomas (Di Majo et al., 1996) and a group called lymphoreticular tumours (Grahn et al., 1992), which includes generalised lymphosarcoma, lymphocytic-lymphoblastic lymphomas, reticulum cell sarcomas, and myeloid leukaemia.

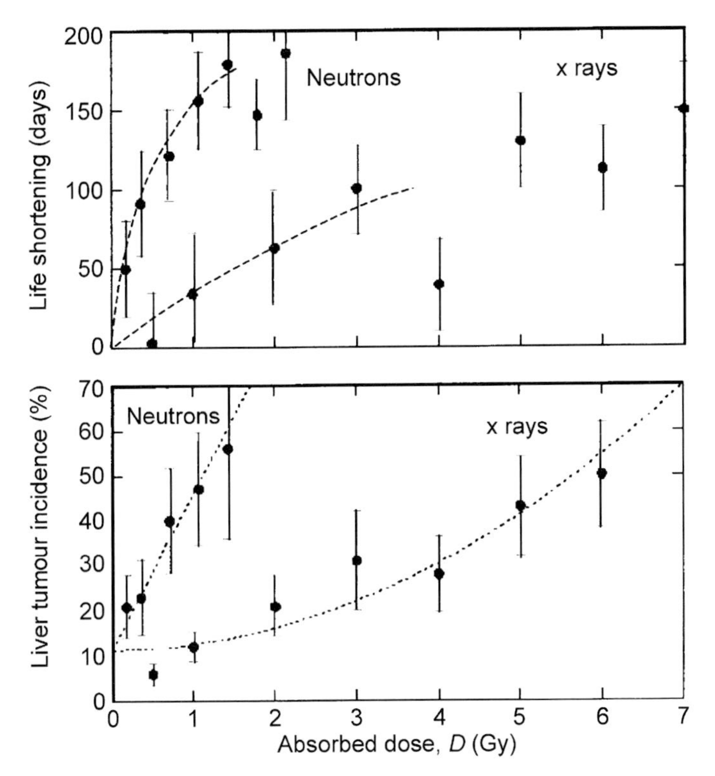

Fig. 3.1. Life shortening in male BC3F1 mice by 250 kV x rays and fission spectrum neutrons (upper panel), and the incidence of liver tumours (lower panel) according to data (and standard errors) by Di Majo et al (1990). The broken lines in the upper panel are for visual guidance only; in the lower panel, they represent the fit reported by the authors to the liver tumour data. According to this fit, the relative biological effectiveness (RBE) of neutrons against an x-ray dose of 2 Gy is about 15. While the RBE values can be stated with some precision for high doses, it is apparent that the RBE at minimal doses (RBE_M) might be larger but cannot be determined from these data. For life shortening, the RBE values appear to be of similar magnitude.

(127) The rates of reticulum cell sarcomas in mice are not increased by radiation. Thymic lymphomas occur predominately in female mice and are readily induced in C57BL, RFM, and RF/UN mice by irradiation, particularly by fractionated high doses. Myeloid leukaemia is predominant in males of certain mouse strains with a very low natural incidence.

(128) Leukaemias are induced in humans by photon radiation with high excess relative risks (ERRs) per unit dose. However, the excess absolute risks (EARs) are substantially smaller than the EARs of radiation-induced solid tumours. The susceptibility for the induction of carcinomas is greater than for sarcomas in humans, and greater attention should, thus, be given to animal data on myeloid leukaemia than to cancers of the lymphoid system, especially thymic lymphoma.

(129) The reported values of RBE for leukaemias and lymphoma in mice are shown in Table 3.3.

Table 3.3. Relative biological effectiveness of neutrons for induction of leukaemias and lymphomas in mice

Mouse strain	Tumour type	Neutron energy	Reference radiation	RBE	Reference
RF/Un	Myeloid leukaemia	Fission (LDR)	γ rays (LDR)	16	Upton et al. (1970)
RFM	Myeloid leukaemia	Fission	γ rays	2.8	Ullrich and Preston (1987)
CBA/CNE	Myeloid leukaemia	Fission	x rays	2.3	Di Majo et al. (1996)
RF/Un	Thymic lymphoma	Fission (LDR)	γ rays (LDR)	3.3	Upton et al. (1970)
RFM	Thymic lymphoma	Fission (LDR)	γ rays (LDR)	27	Ullrich et al. (1976)
CBA/CNE	Malignant lymphoma	Fission	x rays	11	Di Majo et al. (1996)

RBE, relative biological effectiveness; LDR, low dose rate.

3.3.3. Rats

(130) Among the difficulties in the analysis of experimental animal data that are used for estimating RBE is the fact that the degree of lethality of the observed tumours varies. The decision whether a tumour is the cause of death or is found incidentally at autopsy influences the choice of the proper analysis. It can be argued that RBE of lethal tumours is more important for extrapolation of values of RBE in the estimate of risk for humans.

(131) Wolf et al. (2000) analysed data from the large study on Sprague-Dawley rats at the French CEA that employed very small single doses of fission neutrons down to 12 mGy, and estimated RBE for tumours that were assessed to have contributed significantly to lethality. RBE was deduced from a comparison of the cumulative hazard functions due to neutrons and γ rays in terms of a non-parametric analysis for a variety of mathematical models. A neutron absorbed dose of 20 mGy and a γ-ray absorbed dose of 1 Gy produced the same substantial enhancement of the rate of 'lethal' solid tumours. Since the experiment covered a broad dose range from 12 mGy of neutrons to 7 Gy of γ rays, the data are given in Fig. 3.2 against a logarithmic dose scale. The effect amounts to an ERR of 1.9 ($\pm$0.6) for 20 mGy of neutrons and 1.9 ($\pm$0.4) for 1 Gy γ rays. The smaller neutron dose of 12 mGy produced an ERR of 1.0 ($\pm$0.5).

(132) The equivalence of 20 mGy of neutrons and 1 Gy of γ rays corresponds to a neutron RBE of 50, which is in good agreement with the earlier estimate of RBE for lung tumours that were considered non-lethal (Lafuma et al., 1989). There were no data for γ-ray doses below 1 Gy, and it is therefore not possible to determine whether RBE_M for the induction of lethal tumours is about 50 or exceeds this value. The data for life shortening in this study suggest, as seen in the lower panel of Fig. 3.2, the somewhat lower RBE of about 30 against 1 Gy of γ rays.

(133) Shellabarger et al. (1980) performed earlier studies on Sprague-Dawley rats with very low neutron doses down to 1 mGy. They reported an RBE of about 50 for 0.43 MeV neutrons relative to 0.28 Gy of 250 kV x rays for the benign mammary fibroadenomas, while RBE was about 15 against 0.85 Gy of x rays. The same group estimated RBE for both fibroadenoma and adenocarcinomas to be about 10, but in

combination with diethylstilbestrol, RBE for adenocarcinomas and all mammary tumours rose to about 100 (Shellabarger et al., 1982). Broerse et al. (1985, 1991) reported an RBE of 7 for 0.5 MeV neutrons vs x rays in a study of mammary fibroadenomas in a different colony of Sprague-Dawley rats than that used by Shellabarger et al. (1980), and an RBE of 15 for adenocarcinomas in WAG/Rij rats.

3.3.4. In-vitro neoplastic transformation

(134) The only estimates of neutron RBE for transformation of primary cells (Borek et al., 1978, 1983; Hall et al., 1982) are for 430 keV neutrons vs 250 kV x rays as the reference radiation in the first two studies and γ rays in the third study. The results indicated that x rays were twice as effective as γ rays. There was no attempt to estimate the maximum RBE, but the RBE values appeared high at the low doses. An increase in neoplastic transformation was detected at 1 mGy of neutrons but the

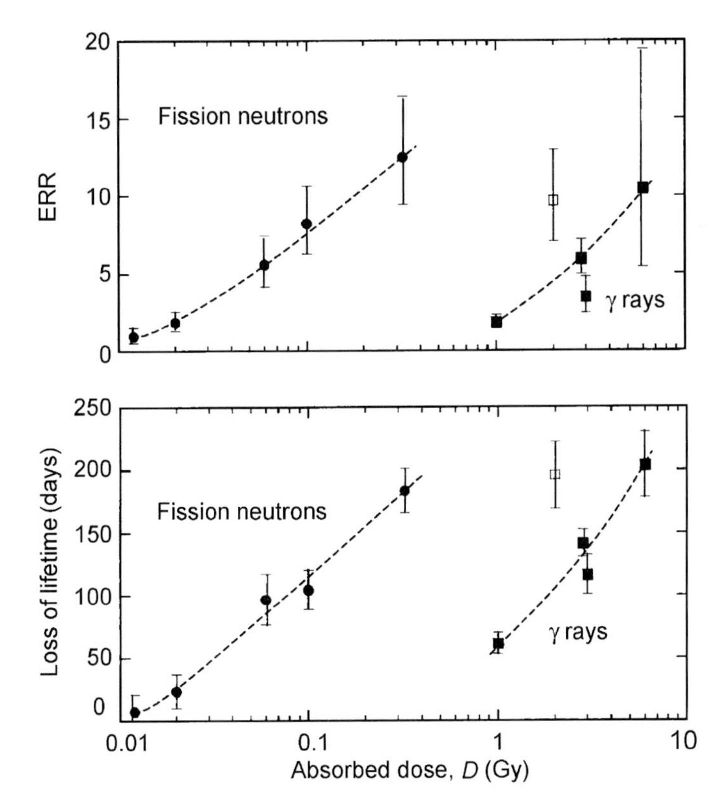

Fig. 3.2. Upper panel: Incidence (and standard errors) of lethal tumours in male Sprague-Dawley rats from the experiments at the French CEA (Wolf et al., 2000). Solid circles, experiments with fission neutrons; solid squares, experiments with γ rays. The open square at 2 Gy belongs to an experiment at ultra-high dose rate with a pulsed x-ray generator. Lower panel: life shortening observed in the same experiment. The broken lines serve for visual guidance only.

confidence limits were broad (Hall et al., 1982). There have been many studies with the C3H10T1/2 cell line. Since these cells are transformed, aneuploid, and unstable, the stage of the neoplastic process affected by irradiation is not clear.

(135) Han and Elkind (1979) found an RBE for fission neutrons vs 250 kV x rays of about 3 at a neutron dose of 3 Gy, and this rose to 10 at 0.2 Gy. In another study, RBE for low-dose-rate neutrons vs low-dose-rate γ rays was 35 (Hill et al., 1984). When both radiations were fractionated, RBE rose to 70 (Hill et al., 1985). Within a range of 0.23–13.7 MeV mono-energetic neutrons, Miller and Hall (1991) observed a modest enhancement with protraction only for 5.9 MeV neutrons. RBE values for 40–350 keV neutrons are shown in Table 3.4.

(136) These results are consistent with predictions based on microdosimetry (Kellerer and Rossi, 1972), and lend support to a lower w_R for neutrons in the 10–100-keV range (ICRP, 1991).

3.3.5. Chromosome aberrations

(137) Apart from the small amount of data for in-vitro cell transformation, the selection of the w_R values for different neutron energies must rely on microdosimetric considerations and RBE values for chromosome aberrations. The report of the Joint ICRP–ICRU Task Group on 'The Quality Factor in Radiation Protection' (ICRU, 1986) has taken special account of the in-vitro data on chromosome aberrations. Prominent among these data were RBE values of neutrons for chromosome aberrations from a systematic study of energy dependence (Edwards et al., 1982) that are shown in Table 3.5.

Table 3.4. Relative biological effectiveness for neutron-induced oncogenic transformation of C3H10T1/2 cells (Miller et al., 2000)

Neutron energy (keV)	RBE[a]
40	3.7±1.9
70	6.6±3.1
350	7.2±3.3

RBE, relative biological effectiveness.

[a] RBE determined from the ratios of the α values for neutrons and 250 kV x rays.

Table 3.5. Relative biological effectiveness for neutron-induced chromosome aberrations in human lymphocytes (Edwards et al., 1982)

Neutron energy (MeV)	RBE[a]
(Fission neutrons) 0.7	53
(Fission neutrons) 0.9	46
(^{252}Cf neutrons) 2.13	38
(Cyclotron neutrons) 7.6	30

RBE, relative biological effectiveness.

[a] RBE values: ratio of α coefficients for neutrons and ^{60}Co γ rays.

(138) Schmid et al. (1998, 2000, 2002a, 2003) reported results for dicentric chromosomes for a broad energy range of mono-energetic neutrons and for blood samples of the same donor. Their data, represented in Fig. 3.3, show that the neutrons reach maximum effectiveness at an energy near 0.4 MeV, a fact that had first been established in a series of major experiments (Bateman et al., 1972; Sparrow et al., 1972; Shellabarger et al., 1974) that figured prominently in the development of microdosimetry by H.H. Rossi. The upper panel in the figure shows the initial slope and the standard error intervals derived in terms of linear-quadratic fits to the dose relationships. For comparison, the lower panel represents RBE against ^{60}Co γ rays (see Fig. 2.1). The results are given for low-dose RBE and also for RBE against 1 Gy γ rays. The bars indicate the standard error intervals. Since the uncertainty of RBE_M values is predominantly due to the uncertainty of the initial slope

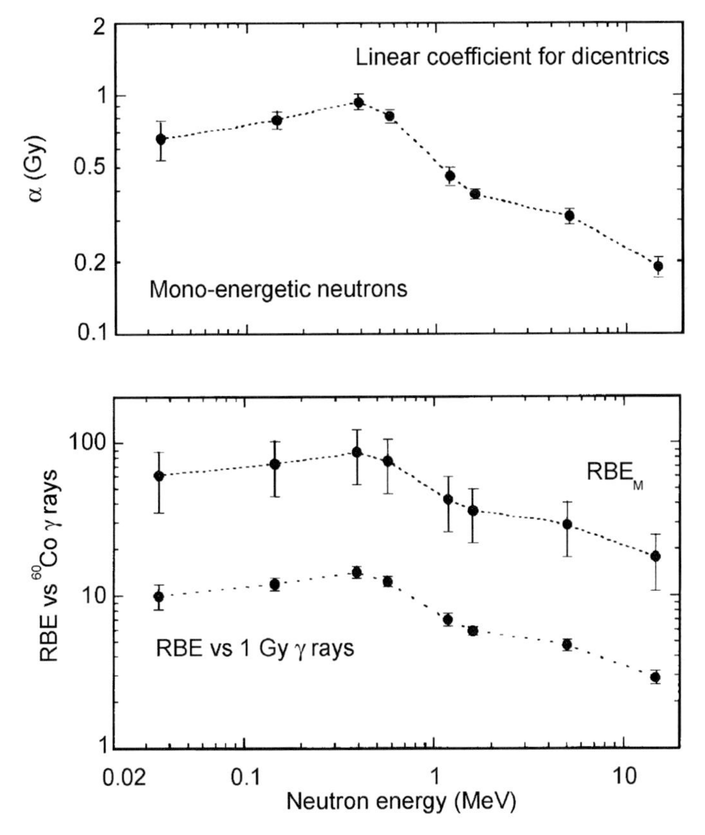

Fig. 3.3. Upper panel: The linear coefficient, α (and standard error), from a linear-quadratic fit to the dose dependencies reported by Schmid et al (1998, 2000, 2002a, 2003) for dicentric chromosome aberrations induced in human lymphocytes by mono-energetic neutrons of different energies. The data point at 1.6 MeV is based on an experiment with fission spectrum neutrons. Lower panel: the relative biological effectiveness at minimal doses (RBE_M, and standard error) vs ^{60}Co γ rays. The curve below gives the high-dose RBE against 1 Gy of γ rays [^{60}Co γ-ray data by Bauchinger et al. (1983) as in Fig. 2.1].

[$\alpha = 0.011(\pm 0.004)/\text{Gy}$] for the γ rays, it is substantially larger than the uncertainty of the neutron data.

(139) The much lower RBE values against the high dose of 1 Gy of γ rays are included in Fig. 3.3 in order to alert one to the strong dependence of RBE on γ-ray dose in the case of the chromosome aberrations with their highly curved dependence on γ-ray dose. RBE against high dose is also of interest because its relative standard error is only slightly larger than that of the neutron data. The reason is that the yield of dicentrics at 1 Gy of γ rays is much more precisely determined than the initial slope.

3.3.6. Mutations

(140) Pink mutations in cells of the stamen hairs of *Tradescantia* were one of the first endpoints that demonstrated, with high experimental precision and down to remarkably low neutron doses of fractions of 1 mGy, the high RBE of 430 keV neutrons against 250 kV x rays (Sparrow et al., 1972). Figure 3.4 gives the dose dependencies that are—apart from the non-linearity at higher doses—well represented by a linear dependence for the neutrons and a linear-quadratic dependence for the x rays. RBE_M has a high value of about 50.

(141) Similarly high or higher values of RBE_M for 430 keV neutrons against x rays were obtained in related studies at Brookhaven National Laboratory in the experiments by Bateman et al. (1972) on lens opacification in mice (see Chapter 5) and, as stated earlier in Section 3.3.3, in some of the experiments on mammary tumours in female Sprague-Dawley rats (Shellabarger et al., 1974, 1980).

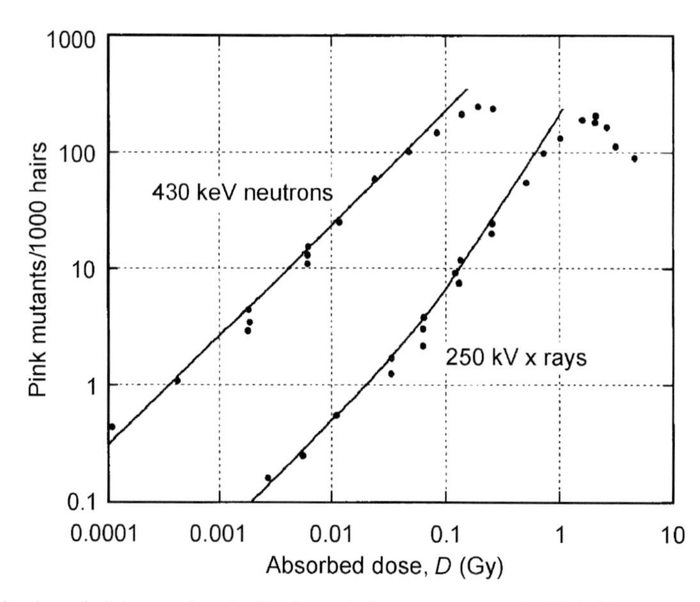

Fig. 3.4. Induction of pink mutations in *Tradescantia* by mono-energetic 430 keV neutrons and 250 kV x rays (Sparrow et al., 1972).

(142) A major conclusion from the series of experiments performed at Brookhaven National Laboratory was that 430 keV is about the most effective neutron energy for the various endpoints that have been studied. This is in agreement with the data on chromosome aberrations in Fig. 3.3. The observation can be readily understood from the fact that a 0.4 MeV neutron transfers, on average, 200 keV to a recoil proton. This is just the right energy for the proton to pass through its Bragg peak and, thus, to deposit, along its short range of about 3 μm, a substantial energy with maximum LET in the nucleus of a cell.

(143) As will be pointed out in the subsequent chapter, the most effective energy of neutrons incident on the human body is about 1 MeV, rather than 0.4 MeV. The predominant reason for this difference lies in the fact that incident neutrons of less than about 1 MeV tend to produce, in addition to the dose from heavy recoils (predominantly protons), a substantial γ-ray dose component in a large receptor, such as the human body (see Section 4.3.1). This dilutes the RBE of the radiation if, as in the definition of the effective dose from neutrons, this γ-ray dose contribution is taken to be part of the 'neutron dose'. With 1 MeV neutrons, the γ-ray dose contribution is much less than with 0.4 MeV neutrons (see Fig. 4.2). No comparable γ contribution occurs in small experimental samples or small animals, and this explains why w_R for neutrons must be smaller than the representative RBE_M values from experimental studies at low neutron energies.

3.4. Protons

(144) *Publication 60* (ICRP, 1991) recommends a $w_R = 5$ for protons with energy greater than 2 MeV. Subsequently, the NCRP—in one of the very rare points of disagreement with the ICRP—argued against the recommendation and proposed a $w_R = 2$ for protons in excess of 2 MeV (NCRP, 1993). The NCRP added the suggestion that a w_R of about 1 would be appropriate for neutron energies above 100 MeV. Neither report details the radiobiological data, or RBE values, on which selection of the w_R values was based.

(145) The choice of method of accounting for the radiation quality of protons for radiation protection purposes has become important in the estimate of effective doses that may be incurred by crews of aircraft flying at high altitudes. The ambient radiation increases with altitude by about 15% (dependent on latitude) for each increase of about 600 m (~2000 ft). At the altitudes of transcontinental flights, about 12 km, the composition and energy spectrum of radiation fields within the aircraft are complex. The radiation is a mixture of primary and secondary high- and low-LET radiations. The cosmic rays, which consist mainly of protons of a broad range of energies, interact with the atmosphere resulting in a cascade of different types of radiation. The most important reaction produces secondary neutrons and protons.

(146) To assess the risk to air crews, it is necessary to determine the effective dose and, therefore, w_R. Different values for the effective dose have been computed by the use of three different w_R values. The different choices were $w_R = 5$ (ICRP, 1991), $w_R = 1$ (NCRP, 1993; recommendation for energies >100 MeV), or, as selected by the Commission of the European Communities, the current Q-LET relationship. If

$w_R = 5$ is used, the annual effective dose can reach a level that, based on the recommendations made in *Publication 60* (ICRP, 1991), would require air crews to be classified as radiation workers.

(147) The recommendations of the Commission of the European Communities include estimation of doses to air crews as part of decision making and control of exposures. This is achieved for protons by calculating the operational quantity ambient dose equivalent, H^*, which uses the Q-LET relationship given by *Publication 60* (ICRP, 1991) and then multiplying the result by 0.8. The choice of H^* as reference quantity reflects the desire to make the reference quantity measurable (see Section 4.1.4) but, more importantly, it is based on the judgement that a w_R value of 5 for protons is too high. The factor of 0.8 was probably chosen to account for the fact that the reference depth in the definition of H^* is too shallow to represent the shielding of the high-energy protons in the body.

(148) There are both biophysical and radiobiological aspects that must be considered in the selection of appropriate w_R values for energetic protons.

3.4.1. Biophysical considerations

(149) Radiobiological measurements of RBE have often been related to LET, and many relationships between RBE and LET have been published. The dependence $Q(L)$ of the quality factor on LET that was introduced in *Publication 26* (ICRP, 1977) and the subsequent modification that was specified in *Publication 60* (ICRP, 1991) (see Fig. 1.1 and Table 1.1) were both judgements based on a review of experimentally measured RBE values extrapolated to low doses. At their respective times of derivation, they were seen to provide a reasonable approximation to the relationship between the increase of cancer risk per unit dose and the LET of a radiation. The calculated average Q and the range in tissue of protons are shown in Table 3.6.

(150) It can be seen that only protons below 4 MeV have a Q of 5 or more. The penetration of tissue at this energy is superficial. At proton energies in excess of 4 MeV, the LET is less than 10 keV/μm, which corresponds, according to the current $Q(L)$ convention to $Q = 1$. Only the track ends below 4 MeV contribute, therefore,

Table 3.6. Average quality factor for protons completely stopped in tissue computed in terms of the stopping powers in ICRU Report 49 (1993a) and the earlier (ICRP, 1977) or the current (ICRP, 1991) numerical values for $Q(L)$

Proton energy (MeV)	ICRP average Q		Range (cm)
	1977	1991	
1	8.8	13.0	0.002
2	6.6	9.0	0.008
4	4.0	4.75	0.025
10	2.1	2.5	0.12
20	1.6	1.75	0.43
50	1.2	1.3	2.2
100	1.1	1.15	7.7
150	1.06	1.1	16

to increased values of Q, which results (with the unit MeV for the proton energy, E_p) in the relationship for the average Q:

$$Q(E_p) = 1 + 15/E_p \text{ for } E_p > 4 \text{ MeV} \quad (3.6)$$

As is readily seen, the same formula applies to the average proton energy if all initial energies exceed 4 MeV and all protons are stopped in the tissue or organ:

$$Q(E_p) = 1 + 15/E_p \quad (3.7)$$

(151) The mean energy of the protons entering an organ is less than the mean energy of the protons incident on the body. The deepest lying organs will, therefore, be associated with the largest value. However, a degradation of the typical spectrum at aviation altitudes will, even for the deeper organs, result in mean energies close to or larger than 100 MeV. The mean Q for the protons will, therefore, not usually be larger than about 1.15.

(152) In Section 4.4.4, it will be pointed out that the effective quality factor, q_E, i.e. the ratio of the effective dose equivalent to the organ-weighted absorbed dose, is somewhat larger at very high neutron energies (≈1 GeV). This is due to secondaries from nuclear interactions of the protons in the body. In view of this added aspect, the w_R value of 2 will be recommended for cosmic ray protons. However, this does not change the conclusion that it is unjustified to assign the current $w_R = 5$ to protons.

3.4.2. Radiobiological data

(153) There are no data from exposures of humans to protons that relate to stochastic effects. The data available from experiments on monkeys, rats, and mice are restricted to a limited number of proton energies and to studies that do not provide values of RBE_M. There is an extensive literature on proton effects on cells in vitro, clonogenic cells, and chromosomes (see reviews by Raju, 1995; Paganetti et al., 1997; Gerwick and Kozin, 1999; Skarsgard, 1998).

(154) The results for lethality, in-vitro and in-vivo cell killing, mutations, chromosome aberrations, and normal tissue effects indicate RBE values less than 2.

Table 3.7. Effects of proton irradiation

Biological effect	Species	Proton energy/MeV	RBE	Reference
Lethality $LD_{50/30}$	Mouse	126	0.7	Ryzkov et al. (1967)[a]
Lethality $LD_{50/30}$	Mouse	50	1.2	Grigoryev et al. (1969)[a]
Lethality	Monkey	32–2300	about 1	Dalrymple et al. (1991)
Cataract	Mouse	50	about 1	Fedorenko et al. (1995)
Cataract	Monkey	160	about 1	Fedorenko et al. (1995)
Cataract	Monkey	55	about 1	Niemer-Tucker et al. (1999)
Gut, lens, and skin	Mouse	160	0.8–1.3	Urano et al. (1984)
Intestinal crypt	Mouse	85	0.9–1.2	Gueulette et al. (1996)

RBE, relative biological effectiveness.
[a] Cited in Tobias and Grigoryev (1975).

Table 3.7 shows RBE values for protons from 50 to 160 MeV in relation to normal tissue effects.

(155) A number of chromosome aberration studies have been performed with different proton energies; however, most if not all have been done without corresponding data for a reference radiation. If the α coefficient for 8.7 MeV protons (Edwards et al., 1985) is compared with the α coefficient for ^{60}Co γ rays from the same laboratory, RBE is about 3.

(156) Life shortening and the risk of cancer were followed in Rhesus monkeys that were part of a study of the acute effects of protons ranging from 32 MeV to 2.3 GeV (Dalrymple et al., 1991). The RBE value for all of the acute-effect studies was 1. Unfortunately, the doses and the number of exposed animals were insufficient for the determination of RBE for the subsequent cancers. Estimates of relative risk based on the incidence of cancer after exposure to x rays and 138 MeV protons were similar and were no higher for protons of the other energies (Wood, 1991). The 32 and 55 MeV protons produced a very non-uniform dose distribution despite rotation of the animals during their exposure.

(157) A study of life shortening and tumour induction by 60 MeV protons and a comparison with exposures to 300 kV x rays in RF/Un female mice was carried out by Clapp et al. (1974). The mice were rotated during exposure and it was estimated that the average proton energy in the body was about 40 MeV and the average LET was about 1.5 keV/μm. In view of this low LET value, it was not surprising that the RBE value for life shortening and tumour induction did not exceed 1.

(158) The clear conclusion from the experimental work is that the effectiveness of high energy protons is roughly the same as that of other low-LET radiations. It follows that a radiation weighting factor $w_R = 5$ is much too high. A $\boldsymbol{w_R = 2}$, i.e., a value larger than 1, is recommended in view of the high-LET secondaries that are produced by high-energy protons in the human body and are absorbed in the body. Experiments with small rodents would not adequately exhibit this effect and any analysis of experimental data for high energy protons must, therefore, be performed in terms of $Q(L)$.

3.5. α particles

(159) Humans are exposed to α rays from internal emitters, such as the progeny of radon (NAS, 1999). There have been extensive studies in humans of the carcinogenic effect of radon, radium, and Thorotrast, a contrast medium that emitted α particles, and of various α-particle-emitting radionuclides in experimental animals (see Radiat. Res., 1999, 152, Suppl.).

(160) There are limitations to the determination of generally applicable RBE or RBE_M values because of the complexity of the dosimetry due to the non-homogeneity of the energy deposition. The range of the α particles in tissues is small, a matter of micrometres, and the precise nature and location of the target cells in some of the most relevant tissues is not known.

(161) An indication of the relative effectiveness for the induction of bone sarcomas has been determined by a comparison of the effect of α and β particles, and estimates of liver and lung cancer have been made. There has been a recent review of many of

the aspects of the effects of α-particle irradiation (see Machinami et al., 1999) but no RBE values were given.

3.5.1. Lung cancer

(162) α particles are emitted by the progeny of radon, a ubiquitous gas that is concentrated in uranium mines and buildings on rock and soil with uranium. Radon has been shown to cause excess lung cancers in uranium miners, and is believed to be a risk in homes with high levels of radon.

(163) There have been two approaches to the estimate of risk of lung cancer from radon exposure:

- an estimate based on epidemiological studies of populations exposed to radon;
- an estimate based on data from the atomic bomb survivors exposed to low-LET radiation, on the dosimetric model for the lung, and on an adjustment for radiation quality in terms of w_R.

In terms of these two approaches, the estimates of ERR of lung cancer per unit organ-equivalent dose differ by a factor of about 4, the higher values resulting from the dosimetric approach.

(164) BEIR VI (NAS, 1999) used the epidemiological approach based on the studies of the uranium miners. In this approach, the lung-equivalent dose and its contribution to effective dose is directly linked to the level of radon exposure and the observed risk per unit exposure. This avoids the use of the dosimetric model that is, as discussed below, still subject to considerable uncertainties. The dosimetric model is, nevertheless, of obvious importance and it has been a matter of discussion for radon exposures. It is, furthermore, of special interest with regard to emerging data on the lung cancer due to plutonium inhalation in the nuclear workers of Mayak (Kreisheimer et al., 2000; Koshurnikova et al., 2002).

(165) The w_R for α particles has been set equal to 20 (ICRP, 1991), but this is merely a rough representation of estimated values of RBE that show considerable variations. Brenner et al. (1995) suggested that Q should be 10. Their statement was based on data for oncogenic transformation in C3H 10T1/2 cells exposed to mono-energetic charged particles of different LET, ranging from 4 to 600 keV/μm; it took into consideration the depth of the assumed target cells in the tissue. The authors pointed out that much of the deposition of energy takes place in the highest LET range of the α particle, and that there is a saturation effect due to overkill. Due to the short α-particle track, it is essential to know the location of the target cells. The depth of penetration by the α particles depends on whether smoking has increased the thickness of the layer of mucus on the surface of the epithelium or has caused hyperplasia. On the assumption that the basal cells of the bronchial epithelium are the target cells and that smoking has increased the mucus layer, the estimated average Q is 10, but it is 13 if the suprabasal cells are susceptible to initiation of cancer. Thus, Brenner et al. (1995) believe that a value of 10 should be used in the derivation of the effective dose. In an estimate of the risk of lung cancer by the dosimetric method, Burchall and James (1994) invoked a quality factor of 20.

(166) The uncertainties in the estimate of the risk of lung cancer from exposure to α particles are, thus, largely a question of the appropriate weighting factor, which reflects the lack of a reliable RBE that relates to the human bronchial epithelium.

(167) While the general pattern of the rate of oncogenic transformation as a function of dose and LET is consistent with other reports, the C3H10T1/2 cell system is very different from the presumptive target cells in the lung. The C3H10T1/2 cells are a heterogeneous, aneuploid population of transformed cells and are usually irradiated while flattened on the surface of a dish. Both the likelihood of a hyperdiploid amount of DNA and the geometry of the cells in the dish suggest that the amount of DNA traversed by a single particle will differ significantly from that in the diploid near-spherical nucleus of the cells of the basal layer of the bronchial epithelium.

(168) These features must have an important influence on the probability of cell killing and inititiation of cancer. There is a significant difference in the relationship of cell killing and the number of particle traversals that depends on the shape of the cell (Ford and Terzaghi, 1993). When the cell is in the form in which it exists in the epithelium, the probability of a particle traversal killing the cell is high, whereas in the dish, it takes multiple α-particle traversals of C3H10T1/2 cells to cause a high probability of malignant transformation (Lloyd et al., 1979).

(169) Cell–cell communication, a characteristic of cells in stratified and pseudo-stratified epithelia, is important in the expression and suppression of initiated cells (Terzaghi and Ford, 1994), and this modifying factor may be expressed differently in a dish of fibroblasts.

(170) The induction of bronchial carcinomas by α particles is one case in which the risk estimate may be influenced through a bystander effect. The risk estimate for radiation-induced cancer in most organs includes any potential bystander effect when the risk is based on organ dose. In the case of α-radiation-induced lung cancer, for which the risk depends on the location of the target cells in relation to the penetration of the α particle, a bystander effect (if it occurs in vivo) would increase the target volume and the number of cells at risk.

(171) The approach of Brenner et al. (1995) illustrates the potential influence of a number of factors that are more important in the induction of bronchogenic carcinoma by α particles than is the case for most other radiation qualities and cancer sites. It is also another example of the necessity of a precise description of the energy and LET of high-LET radiations in the target cell. There is, furthermore, the need for information on the shape of the nuclei of different types of target cells, and on the way the shape determines the probability of cell killing by particle traversals. It is important to establish whether bystander effects exist and how they differ in various types of cell populations. Presumably, if bystander effects do occur in vivo, their effect will vary among cell populations depending on the cell–cell communication.

(172) Lafuma et al. (1989) investigated the comparative effectiveness of low doses of α rays, fission neutrons, and γ rays in the induction of lung cancer in Sprague-Dawley rats, and determined an equivalence ratio of about 15 working level months (WLM) of radon daughters to 10 mGy of fission neutrons (as discussed in Section 3.3.3, RBE of neutrons was about 50 compared with 1 Gy of γ rays). Lundgren et al. (1995) studied the carcinogenicity of repeated inhalation of aerosols of $^{239}PuO_2$ in

rats; when comparing the effect with that of β particles, they estimated RBE to be about 21.

3.5.2. Bone sarcomas - RBE of α rays vs β rays

(173) Mays and Finkel (1980) compared the induction of osteogenic sarcomas by ^{226}Ra and ^{90}Sr in beagles and CF_1 mice, and Lloyd et al. (1994) refined the dosimetry. The RBE value of α rays against β rays was found to be about 3 at the doses that resulted in the highest incidences. The highest RBE of about 25 was observed at the lowest incidence level that could be assessed (8%); the average skeletal dose in the beagles was about 1.1 Gy from the α particles and 27 Gy from the β particles. Raabe et al. (1983) conducted a similar experiment, except that strontium was given in the diet. The RBE value increased from about 9 to about 35 as the incidence of bone sarcomas decreased to 3%. Whether RBE is higher at lower doses is difficult to establish experimentally. The increase in RBE was accounted for by the decrease in the effectiveness of the β particles at low doses: another example of the importance of establishing the effectiveness of the reference radiation.

(174) The calculated dose due to the α particles depends on the location of the target cells. Gössner (1999, 2003) and Gössner et al. (2000) reported that the histogenesis of the fibroblastic-fibrohistiocytic type of bone tumour, which is commonly induced by radiation, involves radiation damage and a disturbance in the remodelling process, which makes it a deterministic effect. If this is the case, a lower RBE might be expected for the fibrosarcomas than the osteogenic sarcomas.

(175) Data for the induction of bone tumours in humans [Evans, 1980; Mays and Spiess, 1984; Stehney, 1995; for the latest reviews, see Fry, 1998; Radiat. Res., 1999, 152 (Suppl.), S1–S171] and experimental data have made it possible to extrapolate the risk of bone cancer induction by plutonium from the risk from radium in terms of the concept of the toxicity ratio[7] which is analogous to RBE.

Table 3.8. Toxicity ratios for specific α-particle emitters (Lloyd et al., 1994)

Radionuclide	Relative toxicity
^{226}Ra	1.0 (reference)
^{224}Ra, single exposure	6±2
^{224}Ra, chronic exposure	16±5
^{228}Ra	2.0±0.5
^{239}Pu, monomeric	16±5
^{239}Pu, polymeric	32±10
^{241}Am	6±0.8
^{228}Th	8.5±2.3

[7] The toxicity ratio (Evans, 1966; Mays et al., 1986) is used to estimate the risk of cancer induction by internal emitters, particularly if no data are available for one specific radionuclide, e.g. plutonium. The use of the toxicity ratio is based on the assumption that the ratio of ^{239}Pu toxicity to ^{226}Ra toxicity in man is approximately equivalent to the ratio of ^{239}Pu toxicity to ^{226}Ra toxicity in experimental animals.

(176) The similarity of the RBE value for ^{226}Ra in the beagle and the mouse gives some support to the validity of extrapolating risk estimates across species.

(177) The relative effectiveness of $^{239}PuO_2$, ^{238}Th, ^{241}Am, ^{228}Ra, and ^{90}Sr for the induction of bone sarcomas was reported by Lloyd et al. (1994), and some of the results are shown in Table 3.8 in terms of the toxicity ratio. The concept of the toxicity ratio has been used to extrapolate the risk of cancer in humans from exposure to plutonium from data obtained in beagles.

(178) It was presumed that the risk coefficient for induction of bone cancer in humans by exposure to low-dose-rate radiation ^{226}Ra of doses less than 10 Gy was 17.1 per 10,000 person Gy average skeletal dose (NCRP, 1991). The risk coefficients in humans for each of the other radionuclides were estimated by multiplying the risk coefficient for ^{226}Ra in man by the toxicity ratio determined in the beagle. For example, the toxicity ratio for monomeric ^{239}Pu to ^{226}Ra in the beagle is 16 and the estimated risk coefficient for bone cancer in humans is 16 times larger than 17.1 per 10,000 person Gy, i.e. it equals 274 per 10,000 person Gy average skeletal dose.

(179) Using the dose to the endosteal cells and a log-normal distribution to represent the uncertain distribution of RBE, a geometric mean of 50 and a geometric standard deviation of 2.8 were obtained for RBE.

(180) Grogan et al. (2001) estimated the risks of cancer induction in four tissues, including bone, in humans from the inhalation of plutonium. The estimates were based on a combination of four approaches, one of which was to modify the risk coefficient obtained from the Life Span Study of the atomic bomb surviors by an RBE for α-particle radiation from plutonium. The relative toxicity is dependent on the dose levels of the radionuclides used in the comparison, and are generally not maximum values. The Pu to Ra toxicity ratio determined in mice (Taylor et al., 1983) was 15.3, while it was 16.5 in the beagle.

3.5.3. Leukaemia

(181) Spiers et al. (1983) reported that the evidence of excess leukaemias in the radium dial painters exposed to ^{226}Ra and ^{228}Ra was inconclusive. In contrast, studies of patients administered Thorotrast reported risk estimates ranging from 40 to 560 cases per 10,000 person Gy. Grogan et al. (2001) used all the available data for both leukaemia mortality and for estimated doses from the Thorotrast patients and the radium dial painters. A Bayesian approach to risk assessment provided the most likely risk estimate of 0.023/Gy. An excess risk of leukaemia, mainly acute myeloid leukaemia and myelodysplastic syndrome, was found in the German Thorotrast patients (van Kaick et al., 1999), the Danish patients (Andersson et al., 1993), and those in Japan (Mori et al., 1999). When an RBE of 20 was used for α particles, the risk estimate of haematological malignancies was lower by a factor of 10 than the estimated risk coefficient for the atomic bomb survivors. Boice (1993) estimated the cumulative risk of leukaemia induced by Thorotrast to be about 1.3 times greater than the estimate from the atomic bomb survivors. Some tentative adjustments were made for the difference in the types of leukaemia in the different populations. It is, of course, difficult to obtain comparable dose estimates but the analysis did not

support a w_R of 20. van Kaick et al. (1999) reported an excess of leukaemias in the studies on the German Thorotrast patients that was comparable with that reported for the Danish patients, but was not 20 times greater per unit equivalent bone marrow dose than the risk in the atomic bomb survivors. The IARC (2001) reviewed all the available data and considered the RBE value for the induction of leukaemia by α particles to lie between 1 and 2.

(182) Breckon and Cox (1990) noted that bone-seeking α-particle emitters are only weakly leukaemogenic, but that it is difficult to determine the sensitivity of the presumptive target cells in the bone marrow. Radionuclides such as ^{224}Ra, ^{226}Ra, and ^{239}Pu tend to concentrate on the surface of the bone. Therefore, many stem cells escape irradiation. In the case of radon, the topography of the distribution is presumably different. The authors assessed the sensitivity of haematopoietic stem cells to α-particle irradiation in the CBA/H mouse that is susceptible to myeloid leukaemia. They compared the relative effectiveness of 250 kV x rays and 3.5 MeV α particles (LET 124 keV/μm) from ^{239}Pu at doses that reduced the stem-cell viability by about a factor of 10. The method that was used made it possible to compare the effect of the α particles and x rays on the survival of the cells and the induction of the re-arrangement on chromosome 2 which is associated with myeloid leukaemia. α-particle irradiation reduced survival a great deal more than it increased the number of cells with the specific chromosome aberration. Based on an indirect estimate, the RBE value of the α particles vs x rays is about 16 for cell killing, but it is about 11 for the number of clones with cells carrying the specific aberration on chromosome 2.

3.5.4. Liver

(183) The studies of the Thorotrast patients, discussed above, and the studies of the Japanese Thorotrast patients provided risk estimates of liver cancer induced by α particles. Grogan et al. (2001) concluded that a value of 20 (geometric mean) with 1.6 geometric standard deviation was the best estimate of an RBE relative to γ rays based on the follow-up of the Thorotrast patients and the atomic bomb survivors.

3.5.5. In-vitro neoplastic transformation

(184) Martin et al. (1995) determined the rate of oncogenic transformation of Syrian hamster embryo cells by ^{4}He particles from an accelerator (RARAF) with LET ranging from 90 to 200 keV/μm, and they determined RBE values with 250 kV x rays as the reference radiation (Table 3.9).

(185) The particles with an LET of 90 keV/μm had the highest RBE for transformation. The maximal rates of transformants per surviving cell were between 0.003 and 0.006, and they were reached at absorbed doses less than 0.05 Gy by the most effective particles, i.e. at LET up to 120 keV/μm. These doses correspond with an average of substantially less than one particle traversal per cell nucleus. The most effective LET for cell inactivation was 120 keV/μm, and at this LET, the dose for 37% survival was 0.22 Gy, which corresponds to an average of about four particle

Table 3.9. Low-dose relative biological effectiveness relative to 250 kV x rays for α-particle-induced cell killing and oncogenic transformation in Syrian hamster embryo cells (Martin et al., 1995)

LET (keV/μm)	RBE_M for cell killing	RBE_M for morphological transformation
90	9	60
100	10	37
120	12	10
150	10	7
180	8	3
200	7	6

RBE_M relative biological effectiveness at minimal doses. LET, linear energy transfer.

traversals per nucleus. A marked decrease in RBE was observed at the higher values of LET for both endpoints.

(186) Riches et al. (1997) exposed human SV40-immortalised aneuploid thyroid epithelial cells to 3.5 MeV α particles from a ^{238}Pu source, and obtained a tentative RBE relative to ^{60}Co γ rays of about 4 for transformation of these cells based on a very limited amount of data. The transformation frequency was assessed in terms of the frequency of tumour induction into athymic nude mice after injection of the irradiated cells.

3.5.6. Chromosome aberrations

(187) Brooks (1975) reported an RBE of 20 for the α emitters ^{239}Pu and ^{241}Am vs protracted irradiation with ^{60}Co γ rays for chromosome aberrations. Against single γ ray doses, they obtained a lower RBE.

(188) The data for the induction of dicentrics in human lymphocytes by α particles and ^{4}He ions of higher energy indicate a larger RBE at the high energy. An RBE of 3 of x rays vs γ rays is assumed in all three investigations, which is in line with the results reported by Edwards et al. (1982), Sasaki et al. (1989) and Schmid et al. (2002a), and is reflected in the difference of RBE values against the two reference radiations in Table 3.10.

Table 3.10. Relative biological effectiveness for the induction of dicentrics in human lymphocytes

	RBE		
α particle (He ion) energy (MeV)	Reference radiation		
	x rays	γ rays	Reference
5.1	8	24	Purrott et al. (1980)
6.1	6	18	Edwards et al. (1980)
23.0	16	48	Takatsuji and Sasaki (1984)

RBE, relative biological effectiveness.

3.6. Heavy ions

(189) Humans are only exposed to significant fluences of heavy ions in space. Heavy ions are a small but important component of galactic cosmic rays. The radiobiology of heavy ions has been studied extensively (Blakely and Kronenberg, 1998), but there are very few investigations that can provide RBE values for carcinogenic effects. Accordingly, there is only a meagre basis for the selection of w_R or Q values. The spectrum of LET of heavy ions is very broad, but the abundance of many of the ions is small and they do not contribute a great deal to the doses that may be incurred in space. Perhaps the most important of the heavy-charged particles is iron, and certainly its effects in tissue are of concern because of its long particle track, because of the high LET, and because of the frequent energetic δ rays that increase the number of cells affected by a single particle traversal.

3.6.1. Tumours

(190) When *Publication 60* (ICRP, 1991) modified the Q-LET relationship that had been previously specified in *Publication 26* (ICRP, 1977), the major change was that the plateau of the curve for Q as a function of LET was replaced by a curve that reached a peak at a Q of 30 (compared with the maximum Q of 20 specified earlier) and then decreased proportional to $L^{-0.5}$ (see Fig. 1.1). This decrease is somewhat less steep than the one that had been proposed by the Joint ICRP–ICRU Task Group on 'The Quality Factor in Radiation Protection' (ICRU, 1986). In the absence of sufficient data for tumour induction, the Joint Task Group had inferred the $1/L$-dependence of the RBE at high LET from studies of endpoints such as chromosome aberrations (Edwards et al., 1982) and mutations (Cox et al., 1977).

(191) Fry et al. (1985) reported RBE values relative to ^{60}Co γ rays for the induction of tumours of the Harderian gland in mice for various high-energy heavy ions. The data, in Table 3.11, were obtained from exposures in the plateau of the beam of the iron ions, and in the spread-out Bragg peaks of the other heavy ion beams. For the iron ions, the LET was about 190 keV/μm. For the other ions, no dose-average LET values were given because the degree of particle fragmentation has not been assessed in the extended Bragg peaks. Muirhead et al. (1997) reviewed data for the induction of tumours in different species and summarised the values of RBE_M. The

Table 3.11. Relative biological effectiveness at minimal dose for the induction of tumours of the Harderian gland in mice by heavy ions

Radiation	Initial energy (MeV/u)	Mean range (g/cm^2)	Depth in beam (g/cm^2)	RBE_M against ^{60}Co γ rays
^{4}Helium	228	26.4	24.3	5
^{12}Carbon	400	22.5	20.3	12
^{20}Neon	425	13.0	10.6	18
^{40}Argon	570	11.0	9.6	27
^{56}Iron	600	11.5	0.5	27

RBE_M, relative biological effectiveness at minimal dose.

range of values for any specific type of tumour was considerable, at least a factor of five in some tumour types.

(192) The estimated RBE_M for the iron ions slightly exceeds the value of 22, specified at an LET of 190 keV/μm by the current $Q(L)$, i.e. the $Q(L)$ introduced in *Publication 60* (ICRP, 1991). For the other ions, the lack of information on LET precludes a direct comparison of RBE_M values to the current convention for $Q(L)$. Nevertheless, it was tentatively concluded from the similarity of the dose–effect relationships for the iron and argon ions that in the RBE vs LET relationship, there may be no sharp peak at 100 keV/μm but rather a plateau at LET values ranging from somewhat less than 100 keV/μm to somewhat greater than 200 keV/μm followed by a decline in effectiveness (NCRP, 2000).

(193) While it is not possible to relate the RBE_M data for the ions other than iron to $Q(L)$, they are informative with regard to w_R which, by definition, includes the dose contribution by fragments created within the body. As far as the exposure depths in the experiments (column 4 in Table 3.11) are roughly comparable to exposure depths in the human body, the results provide guidance on the magnitude of the w_R values to be employed in those situations where the concept of w_R is deemed to be usefully applicable to energetic ions.

(194) For high-energy particles, it is essential—as will be considered in more detail in Chapter 4—to note the difference between $Q(L)$ and w_R. $Q(L)$ is applied to the absorbed dose at a point, i.e. it serves as an internal weighting factor that depends only on the LET of the particles at that point. In contrast, w_R must reflect the entire LET spectrum of the particle and its secondaries created within the body, and it can be substantially different from $Q(L)$ for high-energy ions. The radiation weighting factor, w_R, has been introduced to simplify dosimetric computations for the radiation fields encountered in the radiological protection of workers or of the general public. It was not specifically designed to simplify calculations in radiation fields of very high energy. The complexity of radiation types and exposure situations in space, as well as the required accuracy limits, thus demand careful consideration of the applicability of w_R and $Q(L)$.

(195) Additional data on Harderian gland tumours were obtained by Alpen et al. (1994). Since the $Q(L)$ relationship at high LET is an issue of particular interest to the present report, these data deserve to be considered in detail. The original publication gives the entire data set in terms of a prevalence-fluence diagram because, as the authors state, the 'striking observation is that, in terms of fluence, all of the high-LET ions with LET values in excess of 100 keV/μm are equally effective for tumour induction'. Equal effectiveness per unit fluence would, of course, imply that the effectiveness per unit dose—and, thus, RBE_M—must be inversely proportional to L at high LET.

(196) To examine the issue further, the same data are replotted in Fig. 3.5 against absorbed dose. The logarithmic scale is chosen to indicate- in terms of the horizontal distances between the curves- their RBEs directly. The stopping powers, as given by the authors and listed below Fig. 3.5, are used in the conversion from fluence, Φ, to absorbed dose, D. With the familiar units, Gy, keV/μm, and particles/μm^2, the relationship is:

$$D = 0.16L\Phi \qquad (3.8)$$

(197) It is apparent from Fig. 3.5 that the 600 MeV iron ions with their stopping power of 193 keV/µm have the highest effectiveness among the various types of high-energy particles in the study. The data for the somewhat more densely ionising 350 MeV iron ions and those for the niobium and lanthanum ions with their very high stopping powers are markedly shifted to higher doses at the same prevalence levels, which implies decreased values of RBE compared with the 600 MeV iron ions.

(198) With the current convention $Q(L)$ (see Fig. 1), $Q(L)$ decreases at high LET with the square root of L, i.e. the decline is taken to be less steep than the $1/L$-dependence inferred by Alpen et al. (1994). Visual inspection of Fig. 3.5 suggests that the $1/L$-dependence may well apply. But without detailed statistical analysis, one cannot exclude the somewhat more moderate decrease that corresponds to inverse proportionality to the square root of L.

(199) The essential conclusion is, thus, that the data are not inconsistent with the type of dependence of $Q(L)$ on L that is currently adopted at high LET. If anything, the decrease of RBE_M at very high LET may be somewhat steeper than $Q(L)$ suggests. However, it must be noted that there is a lack of data in the LET range 50–150 keV/µm, and that there is evident need to extend the observations and to determine the RBE–LET relationship in other tumour systems.

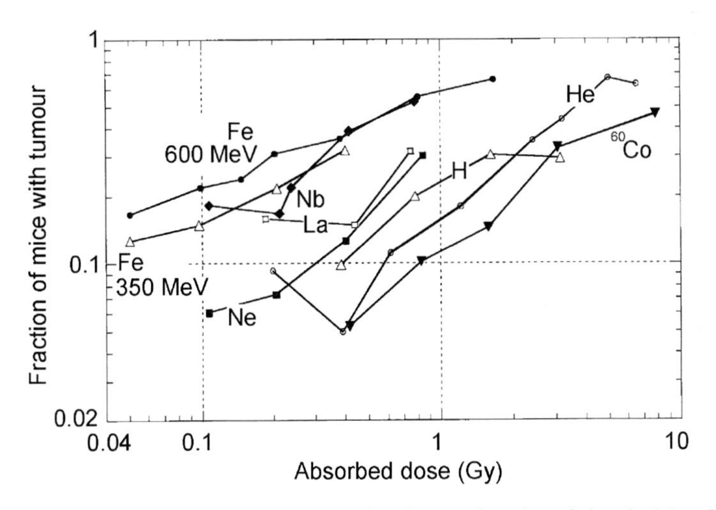

Fig. 3.5. The prevalence of Harderian gland tumours in mice as a function of absorbed dose [from data in Alpen et al. (1994)]. Energies per nucleon and unrestricted linear energy transfer (LET) are listed in the table:

	^{60}Co*	H	He	Ne	Fe	Fe	Nb	La
Energy (MeV)		250	228	670	600	350	600	593
LET (keV/µm)	0.23	0.4	1.6	25	193	253	464	953

*The value for ^{60}Co γ rays is the freqency average LET.

(200) Figure 3.5 indicates, even at the relatively high prevalence level between 0.1 and 0.2, an RBE for the 600 MeV iron ions against ^{60}Co γ rays of about 35. Whether RBE_M is substantially larger cannot be inferred from the data without specific assumptions on the shape of the dose–prevalence relationship.

(201) In this context, it is necessary to note the limitations of linear energy as a parameter of radiation quality. For heavy ions, the LET does not determine the effectiveness uniquely. Of two different heavy ions with the same LET, the one with less charge and less energy per nucleon can be substantially less effective because its track is more narrow and a larger fraction of energy is wasted—due to saturation—in the track core. However, this aspect is not very important for heavy ions with sufficient energy to have significant range. In the experiments by Alpen et al. (1994), three of the high-LET particles had roughly the same high energy per nucleon of about 600 MeV/u. Even the 350 MeV/u iron ions still had a very wide track (about 0.7 MeV maximum δ-ray energy). For incident heavy ions of practical concern, the LET remains, therefore, a meaningful parameter of radiation quality.

(202) Burns and Albert (1981) and Burns et al. (1989) studied the effect of ^{40}Ar (about 125 keV/μm) on the induction of tumours in the skin of rats, and Burns et al. (2001) studied the effects of ^{56}Fe (1 GeV/amu). An upper bound for RBE_M cannot be determined because of the nature of the dose–response curve for electrons, i.e. the reference radiation, which appears to be approximately quadratic in dose. As the dose decreases, RBE appears to increase, but the statistical uncertainty precludes an exact numerical estimate. This underlines the impact of the selection of the reference radiation and the need for a direct method for the determination of the carcinogenic effect of radiations of different qualities.

(203) The available data do not support the adoption of one single w_R for heavy ions. Since w_R values for heavy ions are required for complex radiation fields in space and a realistic assessment is desirable, $Q(L)$ values are more appropriate. Before a dependence on LET that is reliably representative for tumour induction by heavy ions can be derived, there will have to be data for various tissues, especially for heavy ions in the LET range 100–400 keV/μm. Only on the basis of such data could it be established whether the RBE–LET relationship for tumour induction is different from the relationship for cellular endpoints, such as cell killing and induction of mutations and chromosome aberrations (Cox et al., 1977; Edwards et al., 1982, 1986; Edwards, 2001; NCRP, 1990).

(204) The selection of one single value of w_R, such as $w_R = 20$, for incident heavy ions may serve as a conservative approximation for application in the usual circumstances of radiation protection where such exposure situations are of little concern. Situations where w_R values for heavy ions become critical occur predominantly in outer space. In the complex radiation fields that prevail there, the choice of a single w_R would be an oversimplification which conflicts with the available data. With regard to such applications—and especially for critical assessments—it will, therefore, be preferable to use a weighting factor value that depends on the LET of the particles in the organs and tissues of interest. However, as emphasised in Chapter 4, it will be essential to ensure consistency with the w_R system.

3.6.2. In-vitro neoplastic transformation

(205) Yang et al. (1985, 1996) reported increasing RBE values for induction of neoplastic transformation in C3H10T1/2 cells for heavy charged particles up to about 10 at an LET of 100–200 keV/μm. The doses of the reference radiation, x rays, were not sufficiently low to allow the determination of a maximum RBE value. In 1986, the same authors reported that lowering the dose rate of the heavy ion exposure increased the neoplastic transformation rate, but the increase was relatively small (Yang et al., 1996).

3.6.3. Chromosome aberrations

(206) For ions with an LET of 10 keV/μm, Geard (1985) found a yield of chromosome aberrations that varied by up to a factor of 4 for different phases of the cell cycle. The highest yield was found in the G_2 phase. The variation in sensitivity declined as LET increased to 80 keV/μm.The induced aberration rate per unit absorbed dose increased (over the range 10–80 keV/μm) four-fold. Edwards (1997) reported the induction of dicentrics by several heavy ions and RBE_M values calculated from the linear coefficients for both x and γ rays (Table 3.12); RBE of the 250 kV x rays vs ^{60}Co γ rays is taken to be 2 in this analysis.

(207) Testard et al. (1996, 1997) and Obe et al. (1997) determined the frequency of chromosome aberrations in lymphocytes of astronauts exposed to mixed radiation fields in space. They found elevated frequencies. Furthermore, they noted an unusual number of metaphases that resembled so-called rogue cells. These cells have multiple lesions and appear to be similar to those reported by Ritter et al. (1992) in their experimental study with heavy ions that suggested aberrations distinct from those induced by low-LET radiation.

3.6.4. Mutations

(208) Kiefer et al. (2001) determined the yield of hypoxanthine-guanine-phosphoribosyl-transferase (HPRT) mutations in V-79 cells for various heavy ions. The data exhibit large statistical variations, but it is apparent from superposition of all data in Fig. 3.6 that there is a maximum RBE relative to 300 kV x rays at around an LET of

Table 3.12. Relative biological effectiveness at minimal doses for heavy charged particle induction of chromosome aberrations

Heavy ion	LET (keV/μm)	RBE_M vs 250 kV x rays	RBE_M vs ^{60}Co γ rays	Reference
^{3}Helium	24	11	22	Edwards et al. (1985)
^{16}Oxygen	49	12–22	25–32	Edwards (1997)
^{12}Carbon	59	15–18	31–37	Edwards (1997)
^{16}Oxygen	67	10	21	Edwards (1997)
^{20}Neon	460	0.1	0.2	Edwards et al. (1994)

RBE_M, relative biological effectiveness at minimal doses; LET, linear energy transfer.

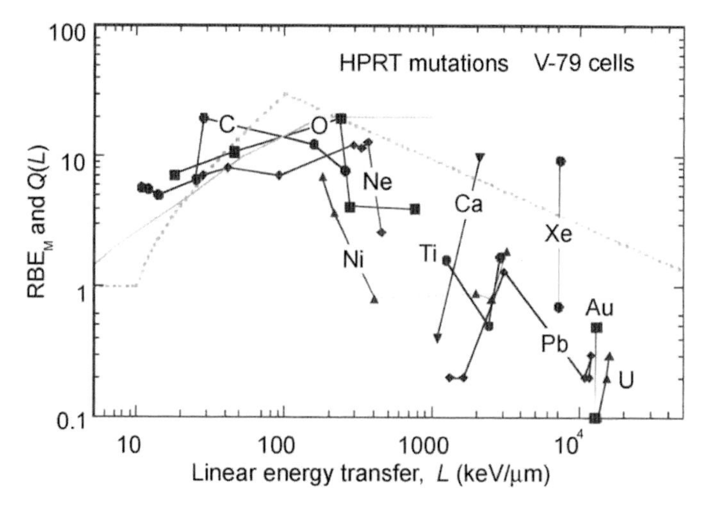

Fig. 3.6. The relative biological effectiveness at minimal doses (RBE_M) relative to 300 kV x rays for the induction of HPRT mutations in V-79 Chinese hamster cells by different heavy ions at different linear energy transfer (LET) (Kiefer et al., 2001). The dotted line represents the current convention for $Q(L)$.

100 keV/μm, and a marked decline of RBE at larger values of LET. The data support, thus, the assumption of a decreasing $Q(L)$ at LET in excess of 100 keV/μm, a conclusion which is in line with a broad spectrum of experimental results obtained over the years at the heavy ion accelerator of the Gesellschaft für Schwerionenforschung (GSI), Darmstadt, Germany, for cell inactivation, chromosome aberrations, and DNA double-strand breaks (Kraft, 1987; Taucher-Scholz and Kraft, 1999).

4. WEIGHTING FACTORS FOR RADIATION QUALITY

(209) The quality factor [$Q(L)$] and the radiation weighting factor (w_R) have been the subject of continued critical discussions with regard to two major aspects. The first aspect is the concept of the quantities and, in particular, the choice of the reference parameters—either linear energy transfer (LET) or kinetic energy and nature of the particle—to represent the radiation quality. The other, largely separate aspect is the choice of appropriate numerical conventions. The two issues are not always distinguished, which has led to difficulties.

(210) Estimates of the late health effects of low radiation exposures and the values of relative biological effectiveness at minimal doses (RBE_M) (see Chapter 3) are subject to considerable uncertainty, and there is thus some discretion in the choice of numerical conventions for radiation protection quantities, including the weighting factors for radiation quality. As the ICRP pointed out in *Publication 60* (ICRP, 1991), no spurious precision should be inferred from the radiation weighting factors. This statement alerts one to the uncertainty in the assessment of the probability of detriment resulting from exposure to radiations of different LET. However, it must not be taken to imply that the definitions of quantities and parameters for use in radiation protection should lack rigour and precision. Quantities for radiation protection practice are, indeed, defined by conventions that are, to some degree, arbitrary. However, once they are defined, they must, in certain circumstances, be assessed by computation and measurement with a degree of precision beyond the accuracy of the underlying radiological information.

(211) The subsequent section deals with the concepts of $Q(L)$ and w_R. It will be seen that difficulties have arisen because the two parameters are, at present, not numerically inter-related. The effective dose, E, is defined in terms of w_R which makes it unsuitable for measurements. $Q(L)$ has, accordingly, been retained in the definition of the operational quantities. However, the use of the two separate concepts impedes comparisons between computations and measurements, and the problem is aggravated because the numerical conventions for w_R and $Q(L)$ are, at present, neither formally related nor, as will be seen, coherent.

(212) With regard to the current parallel use of the 'primary immeasurable quantities' E and H_T and the 'operational quantities' H^* and H_p, Ralph Thomas in Thomas and Lindell (2001), emphasised that 'perhaps the most unfortunate feature of this dual system of quantities is its instability'. To preserve continuity, the existing system of quantities must be retained as far as possible, but it must be made coherent:

> '.. given all the burdens placed on the primary quantities, what possible harm could result if they were precisely defined and in such a manner that they complied with the laws of physics and were mathematically well-behaved? Our understanding of the biology does not forbid it.'

4.1. Issues relating to the concepts of w_R and Q

4.1.1. Origin of the current choice of $Q(L)$

(213) Before an option is developed to remove the current incoherence between w_R and $Q(L)$ for the important case of neutron exposures, it is helpful to retrace the origin of the present numerical convention for $Q(L)$. The transition from $Q(L)$ to the numerical values of w_R will then be explored in Section 4.3.

(214) Figure 1.1 compares the former convention for the quality factor as introduced in *Publication 26* (ICRP, 1977) with the current convention, $Q(L)$, specified in *Publication 60* (ICRP, 1991). The synopsis in Chapter 3 suggests that there is, apart from some uncertainty in the extrapolation to the low-dose-limit RBE_M, reasonable agreement of the experimentally determined values of RBE with the functional dependence $Q(L)$.

(215) The steps that led to the current convention for $Q(L)$ were not put forward in *Publication 60* (ICRP, 1991). However, they can be retraced and it can be seen that the current convention is based on the recommendation in the report given by the Joint ICRP–ICRU Task Group on 'The Quality Factor in Radiation Protection' (ICRU, 1986). The Joint Task Group assessed the experimental data with emphasis on results for chromosome aberrations and with consideration of other relevant radiobiological evidence. Instead of using the accustomed quantity LET as the reference parameter, they used its closely related microdosimetric analogue, the lineal energy, y, and they recommended the relation $Q(y)$ that is shown in the lower panel of Fig. 4.1.

(216) The parameter y was chosen by the Joint Task Group because its distribution can be measured in all types of radiation field. It is defined as the energy lost by a charged particle within a spherical tissue region, typically of 1-μm diameter, divided by the mean chord length of the region which equals two-thirds of the diameter (ICRU, 1983; Rossi and Zaider, 1996). Unrestricted LET, L, which is a mean value, and lineal energy, y, which represents an actual energy deposition, are closely related, and their difference can be largely disregarded for densely ionising radiation. For low-LET radiation, the two concepts differ appreciably and y can have smaller values than LET. Integration of $Q(y)$ over the microdosimetric spectra provides a value of the quality factor of about 0.5 for γ rays, but a value of about 1 for conventional x rays (ICRU, 1986).

(217) While the measurability of y is attractive, the accustomed parameter LET is more convenient in computations, and the reference parameter LET was, therefore, retained in the current system of radiation protection quantities. The conversion of the recommended relation $Q(y)$ into an equivalent LET dependence required microdosimetric considerations which provided the dependence $Q(L)$ which is represented as a broken line in the upper panel of Fig. 4.1 (Kellerer and Hahn, 1988 a,b). ICRP decided to simplify this relation by assigning $Q(L)=1$ to all low-LET radiations ($L < 10$ keV/μm), and by assigning somewhat more conservative $Q(L)$ values to high-LET values as they occur with heavy ions. In spite of these modifications, it is apparent that the current $Q(L)$ is largely in line with the recommendation in the report of the Joint ICRP–ICRU Task Group.

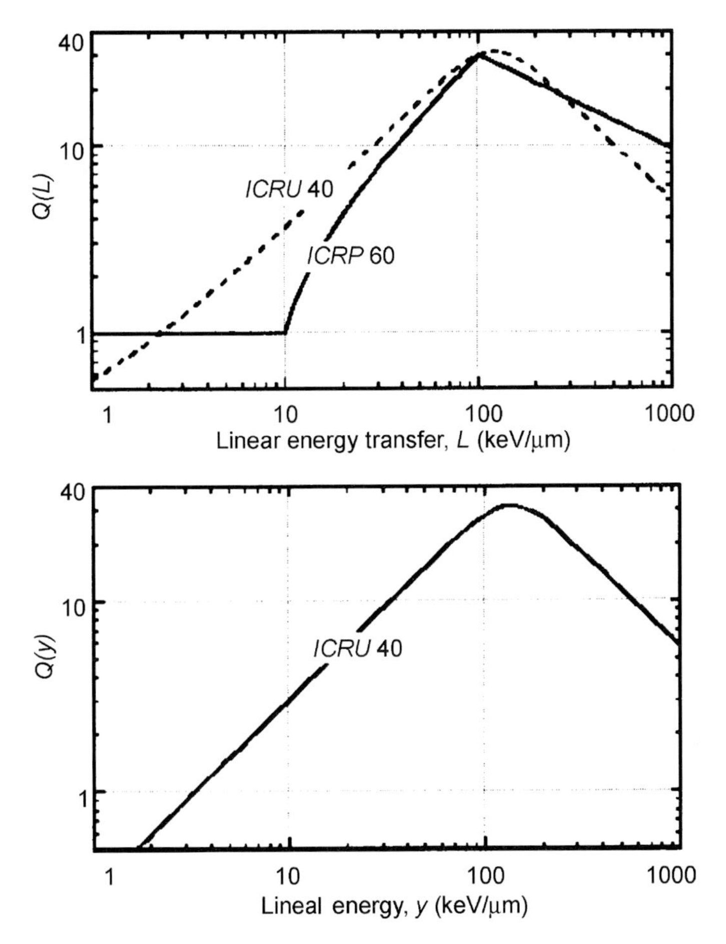

Fig. 4.1. Lower panel: The dependence, $Q(y)$, of the quality factor on the microdosimetric parameter y as proposed by the Joint ICRP–ICRU Task Group (ICRU, 1986). Upper panel: The dependence of the quality factor, $Q(L)$, on linear energy transfer that corresponds to $Q(y)$ (Kellerer and Hahn, 1988a,b) (broken curve) and the current quality factor as introduced by *Publication 60* (ICRP, 1991).

(218) In the subsequent sections, specifically in Section 4.1.3, it will be explained that w_R has been introduced as a simplification that substitutes for $Q(L)$ but was intended to be largely consistent with $Q(L)$. The current w_R is, thus, essentially based on the evaluation of radiobiological data as presented by Sinclair (1982, 1985), NCRP (1990), and the Joint ICRP–ICRU Task Group in 'The Quality Factor in Radiation Protection' (ICRU, 1986).

4.1.2. The need for computation and measurement in radiation protection

(219) Both computations and measurements are required in radiation protection practice. There are many situations where computations alone are sufficient, and

these tend to be performed with increasing facility as computing power keeps increasing. Nevertheless, measurements remain indispensable. The most evident but not the only example is the determination of dose-equivalent quantities in an unknown radiation field. In the following text, the term 'measurement' is understood in the general sense of a determination that may combine measurements with a substantial computational element.

(220) It has occasionally been suggested that the ICRP needs to be concerned exclusively with basic data and computations, while the ICRU must be concerned with the appropriate measurements. However, the two domains of computation and measurement are inseparable, and the basic quantities for radiation protection need to be defined in such a way that measurements to assess compliance with the dose limits are feasible, at least in principle.

(221) The current concept of w_R lacks formal linkage to $Q(L)$. This has introduced the major barrier between computation and measurement which has led to continued problems and to some criticism (Rossi, 1995; Thomas, 2001; Thomas et al., 2002). A modification is, therefore, required to remove the problem with minimal departure from the present system and with maximal attainable coherence.

4.1.3. A peculiar feature in the definition of w_R

(222) The previous reference quantities for radiation protection were the effective dose equivalent, H_E, and the organ dose equivalents, H_T, both being defined in terms of $Q(L)$:

$$H_T = \int_m \int_L Q(L) D_L \, dL \, dm/m \quad \text{and} \quad H_E = \Sigma_T w_T H_T \tag{4.1}$$

where D_L is the distribution of absorbed dose in unrestricted LET, and the integrals range over LET and the mass, m, of the organ.

(223) As pointed out, in line with the recommendations of the Joint ICRP–ICRU Task Group (ICRU, 1986), the ICRP has modified the earlier relationship (ICRP, 1977) between $Q(L)$ and L to reflect higher RBE values for intermediate-energy neutrons and the reduced effectiveness of heavy ions with L greater than 100 keV/μm. There has also been a change in w_T. All subsequent numerical considerations of the former quantities H_E and H_T will relate to the current convention for $Q(L)$ and the current values of w_T.

(224) The effective dose, E, has replaced the former quantity effective dose equivalent, H_E. As stated earlier, it is defined as:

$$E = \Sigma_T w_T H_T \quad \text{with} \quad H_T = \Sigma_T w_R D_{T,R} \tag{4.2}$$

The former quantity, organ dose equivalent, has been correspondingly changed and is now termed the 'organ-equivalent dose'.

(225) In the former quantities, both the absorbed dose and $Q(L)$ were related to the radiation field in the organs. In the new definition [Eq(4.2)], there is an uncommon

element. The absorbed dose, $D_{T,R}$, depends, as in the earlier definition, on the internal field, while w_R depends on the external field with no specification of an equivalent factor that depends on the LET distribution, or possibly another characteristic, of the internal field. This uncommon 'bilocality' makes determinations of E and H_T in mixed radiation fields (e.g. neutron-γ fields) difficult. It requires a quantitative subdivision of the absorbed dose to the organs into each separate contribution, $D_{T,R,}$ that is traced back to a particular component of the external radiation and which must, accordingly, be multiplied by the w_R value assigned to the component. This subdivision cannot be made in terms of the properties of the internal field, i.e. it cannot be achieved through measurements, and while w_R was meant to simplify rough computations, it tends to complicate the more precise computations that are common in radiation protection dosimetry.

(226) Before the weighting factors for radiation quality are further discussed, a question concerning the computation and measurement of effective dose needs to be considered. This is whether E is merely a personal quantity, or whether it can be used also as a quantity defined in terms of a phantom. Primarily, E is related to the body of a specific person whose exposure is quantified. Depending on the type of radiation, the numerical value of E can then vary substantially with the orientation, but especially with the size of the person. Nevertheless, as is the case with many other quantities, the use of E depends on circumstances. Rough approximations, for example in terms of ambient dose equivalent, H^*, will be adequate in some cases. Anthropomorphic standard phantoms will provide closer estimates, and in certain critical assessments special phantoms may be used for computations or measurements that relate to a specific group of persons, or even to an individual. This ambivalence is acceptable and permits flexible application; it is also reflected in the fact that although E is related to a particular person, it is not sex specific, which means, for example, that a sex-averaged tissue weighting factor is assigned to breast, although twice this value would actually apply to females and a value of (roughly) zero to males.

4.1.4. Computation of E or H_T

(227) In computations, the 'bilocality' of H_T does not present a problem. However, it needs to be noted that, while it facilitates rough estimates, the introduction of w_R does not simplify the computation. In fact, the need for the separate quantitative assessment of the absorbed dose contributions that originate from the different components of the external field can complicate the calculation unnecessarily with the current definition. The example of the equilibrium radiation field at aviation altitudes shows this dilemma in an interesting way. However, this issue is not critical and, as far as computations are concerned, the current definition causes no major difficulty.

(228) The situation is different with regard to measurements of E and H_T which would need to be highly complex. In fact, they would require a complete characterisation of the external field, including the directional distribution of all fluences which would then permit computation of E for a given orientation and geometry of

the human body or an appropriate phantom. Such a specification is difficult to attain even in a uniform field; in non-uniform fields that can arise in critical radiation protection situations, it may be virtually impossible. It is for this reason that, in slight exaggeration, E and H_T have been termed 'immeasurable'.

(229) To illustrate the problem, assume that an idealised instrument was available to measure the particle fluence differential in type of radiation and energy outside a phantom and at any point inside the phantom. Such an instrument is not entirely hypothetical, since tissue-equivalent proportional counters can be quite small and can assess absorbed dose as well as radiation quality. One might thus obtain, at least in principle, full information on the radiation types and fluences inside and outside the phantom. This would, of course, include information about the absorbed doses in the organs. However, whenever the external radiation field is composed of types and energies with different values of w_R, there is still no way to determine E or H_T to an organ from the measured data. The reason is that the separate quantitative attribution of the absorbed dose components to the radiation types, R, for the purpose of the radiation weighting, cannot be obtained from the characteristics of the internal field. It needs to be derived from the external field and its degradation in the actual geometry of the exposed body. The remarkable conclusion is that E or H_T cannot be determined even with the combined external and internal measurements and the idealised instrument.

4.1.5. Consequences

(230) Accurate determinations may seem to be an academic issue in radiation protection. Under routine circumstances, where exposures are substantially below the limits, this is indeed the case. There is then no need for accurate assessments. However, in radiological protection, as in other formally adopted and legally binding protection or safety systems, a limit must also be a rigorously defined quantity because exposures must, in certain critical cases, be assessed accurately. Looseness that can involve uncertainties by a factor of 2 or more is tolerable under many routine conditions, but it will make the system inoperable in exactly those critical instances where compliance with regulatory limits is in question and must be reliably quantified.

(231) Due to the peculiarity of its definition, w_R is inapplicable in measurements. Consequently, $Q(L)$, although not formally related to w_R, had to be retained for this purpose and is still used in the operational quantities, ambient dose equivalent, H^*, and personal dose equivalent, H_p. Although the operational quantities are meant to substitute for the regulatory quantity E in routine radiation protection practice, they are now set apart conceptually from E by being linked to the $Q(L)$ system which is not formally related to w_R and E. Fortuitously, the numerical values of H^* and H_p are sufficiently conservative in most exposure situations to serve as an adequate substitute of E for monitoring purposes. Nevertheless, there is a requirement for conceptual clarification and, as will be seen, for some modification of numerical values.

(232) It is important to note that the current problem is not due to the fact that $Q(L)$, which depends on LET, has been replaced by w_R, which depends on radiation type and energy. This change alone would have caused no problem if w_R and $Q(L)$

were traceably inter-related, i.e. if the conventions for w_R and $Q(L)$ were coherent and convertible. Difficulties have arisen because w_R and $Q(L)$ have not been inter-related beyond the vague statement that they are 'broadly compatible'. A practicable formal inter-relation would avoid these problems.

4.2. Effective dose: field or receptor quantity?

(233) While it will be desirable to depart no more than necessary from the current system, it is nevertheless instructive to briefly consider a broader range of options for modifying the current definitions. This section deals primarily with the concepts rather than the numbers. By a suitable system of definitions, the selection of adequate numerical values will, of course, be simplified. The aim is to avoid unrelated conventions for w_R- and LET-dependent weighting factors, such as $Q(L)$. As stated, the definitions of E and H_T refer partly to the external and partly to the internal field in the current system. If this 'bilocality' is to be avoided, a choice needs to be made regarding whether the basic dose-equivalent quantity is defined as a field quantity or a receptor quantity, i.e. whether it is linked to the external field at a point or to the internal field in the body.

4.2.1. Reference to the external field

(234) There are various possible definitions of dose quantities that refer to an external radiation field. Some of the options are:

- $\Sigma\ \Phi_R\ w_R$, i.e. a weighted sum over the fluences due to different radiation components. Here and in the following text, w_R is a weighting factor as in the current definition of E, i.e. it is a factor that depends on radiation type and energy of the fluence component. The numerical values of w_R would differ, of course, from those in the current definition of E.
- $\Sigma\ K_R\ w_R$, i.e. a weighted sum over the contributions from different radiation components to the tissue kerma free in air. w_R is again a weighting factor as in the current definition, but with different numerical values.
- $\Sigma\ D_R\ w_R$, i.e. a weighted sum over the contributions from different radiation components to the dose at a specified location in a specified receptor geometry with the additional, although somewhat artificial, constraint that there is no directional dependence. The numerical values of w_R are again different. A special case thus obtained is the ambient dose equivalent H^*.

(235) The above choices are essentially equivalent, i.e. there is a one-to-one relationship between sets of weighting factors that would be equivalent. In other words, if one option is chosen, the corresponding weighting factors for the other two options can be readily derived. The third option invokes, of course, a specified receptor geometry, but the value of the quantity is determined, through computations—at any point in the receptor free field. This corresponds to the property of tissue kerma (see option 2 above) which is also defined, through computations, for

any point in the receptor free field. These and various similar quantities were considered in the ICRU document (ICRU, 1985) that introduced the current operational quantities. They are noted here to indicate the range of existing options.

(236) If radiation protection were to deal exclusively with penetrating external whole-body exposures, one of the above options might be suitable as the primary dose-equivalent quantity. The adoption of suitably conservative weighting factors could make it acceptable to disregard the directional distribution of the radiation.

(237) There is, however, an important argument against referring the primary dose-equivalent quantity to the external field. The desirability of being able to use the same quantity for whole-body exposures and partial-body exposures, and for external exposure as well as exposure from internal emitters, has been the major rationale for the introduction of the effective dose equivalent. The effective dose equivalent, the predecessor of the effective dose, was defined in terms of the quality factor and it had the desired generality to account for all different geometries and to be equally applicable to external and internal fields. When the current w_R was introduced and the effective dose equivalent was changed to the effective dose, the principal aim continued to be the use of the same quantity for the various exposure situations. This aim appears to rule out the above field quantities as the basic reference quantity for radiation protection.

4.2.2. Reference to the internal field

(238) When the earlier 'quantity effective dose equivalent' was modified and renamed 'effective dose', it was clear that the major aim was simplification, but the specific reasons for the change were not elaborated. If w_R had, at this point, been related to the internal field, it would have been straightforward to determine, through measurements and/or computations, the organ-absorbed doses contributed by the different radiation types and to apply the corresponding weighting factors. If, furthermore, w_R had been chosen to be in line with $Q(L)$, the option to use $Q(L)$ in measurements with tissue-equivalent proportional counters would have been retained, while determinations in terms of radiation type and energy would have been equally admissible.

(239) The linkage of w_R to the 'external' field leads, as has been explained, to difficulties because there is no equivalent LET-dependent weighting factor for the purpose of measurements. As will be explained below, w_R was meant to be coherent with $Q(L)$; this meant that the newly introduced E would have had the same values as H_E if computed with the newly introduced $Q(L)$. However, certain required numerical values were not available at the time, and accordingly the intended aim has not been attained with the current numerical convention w_R. There can be no LET-dependent parameter that is strictly consistent with the current w_R, yet such a parameter is required for measurements. However, in the subsequent sections, an LET-dependent weighting factor will be identified that is equivalent to a moderately modified w_R. It will be seen that this weighting factor is closely related to $Q(L)$. The present w_R and the current definitions of E and H_T can then be retained as an adequate specification under usual circumstances. However, in certain circumstances

where w_R may be inapplicable, or where measurements are part of the determination of E and H_T, the LET-dependent weighting factor can be invoked. This procedure preserves the current concepts and, in essence, the current numerical values, but a specification is added that bridges the gap between computations and measurements.

4.3. Neutrons

(240) The potential modification of the present convention is important primarily for neutrons. The case of high-energy protons also needs to be considered. However, it is a simpler and more obvious issue that involves merely the choice of realistic RBE values (see Chapter 3). The weighting factor for neutrons will be dealt with primarily under the aspect of a modified convention that will essentially preserve the current values of the effective doses from neutrons.

4.3.1. The two components of the neutron dose

(241) With neutrons, the difference between the external and the internal field can be substantial for a large receptor, such as the human body. One reason is degradation of the neutron energy within the body. Another, even more important, reason is that thermal neutron capture generates a substantial component of γ rays in the body. The resulting large difference between the external and the internal radiation field implies a considerable numerical difference between radiation weighting factors that are related to the external field and radiation weighting factors for the internal field.

(242) If a small specimen of tissue is irradiated by neutrons, the absorbed dose is mostly due to recoil protons or heavier recoils from neutron collisions or neutron-induced nuclear reactions. For fast neutron radiation fields, the neutron moderation in a small specimen will be minor: hence, only very few photons from thermal neutron capture are generated and absorbed in the specimen. The absorbed dose, D, in the tissue due to the external neutron field is, therefore, nearly equal to the absorbed dose, D_n, from the charged neutron recoil particles.

(243) When larger tissue volumes, e.g. mice, rats, or humans, are irradiated, the situation becomes increasingly more complex because of the mixed radiation field in the body.

- Neutrons are scattered in the body and partially moderated. The neutron field in the body, therefore, differs from the primary field. The charged particle dose is induced partially by primary neutrons and partially by scattered and moderated neutrons.
- Secondary photons are produced mainly by the H(n,γ)D reaction and also in the decay of excited nuclei from neutron-induced nuclear reactions. Neutron moderation and the relative contribution of secondary photons increase with the size of the receptor volume and with decreasing energy of the neutrons.

(244) The mean absorbed dose, D, by the body from external neutrons can, thus, be described by the sum of D_n, the 'charged heavy particle' dose ('genuine neutron

dose'), and D_γ, the dose from photons which are released in the body by the neutrons:

$$D = D_n + D_\gamma \quad (4.3)$$

(245) Table 4.1 presents values of the relative contribution of the two dose components for a mouse, rat, and an anthropomorphic phantom (Dietze and Siebert, 1994). The mice were simulated by a 3.6-cm-diameter tissue sphere (weight: 25 g) and the rats by a 7.8-cm-diameter sphere (weight: 250 g). Since the calculations for mice and rats were performed in terms of spherical phantoms with typical masses of these animals, the data are approximations. The real shapes may result in somewhat lower contributions of secondary photons, but the essential point is that D_γ is small for mice and rats.

(246) For the anthropomorphic phantom, the organ-averaged absorbed doses were calculated from the mean organ and tissue doses, D_T, by the equation which accounts for the different weighting of the organs:

$$D' = \Sigma_T w_T (D_{T,n} + D_{T,\gamma}) = D'_n + D'_\gamma \quad (4.4)$$

(247) There is, of course, also a dose contribution from photons that are induced by the neutrons outside the irradiated animal or the human body. In experiments, this contribution can, for example, arise in the small containers (e.g. lucite cylinders) in which the small animals are kept during the exposures. It is understood that such 'external' photon contributions are not counted in the total absorbed dose from the neutrons.

(248) The values in Table 4.1 confirm that the photon component, D_γ, is sufficiently small to be disregarded if rodents are exposed to neutrons with energy in

Table 4.1. Fraction of the neutron absorbed dose due to heavy particles (D_n/D) and photons induced in the body by neutrons (D_γ/D) (Dietze and Siebert, 1994). The data for mice and rats relate to spherical phantoms. The data for the anthropomorphic phantom relate to anterior–posterior exposure

E_n	Mice		Rats		Anthropmorphic phantom	
MeV	D_n/D	D_γ/D	D_n/D	D_γ/D	D'_n/D'	D'_γ/D'
Thermal	0.462	0.538	0.293	0.707	0.100	0.900
0.001	0.768	0.232	0.329	0.671	0.098	0.902
0.005	0.956	0.044	0.518	0.482	0.116	0.884
0.01	0.982	0.018	0.661	0.339	0.138	0.862
0.02	0.993	0.007	0.793	0.207	0.170	0.830
0.05	0.997	0.003	0.914	0.086	0.265	0.735
0.1	0.999	0.001	0.960	0.040	0.394	0.606
0.2	1.000	0.000	0.982	0.018	0.540	0.460
0.5	1.000	0.000	0.994	0.006	0.768	0.232
1.0	1.000	0.000	0.998	0.002	0.846	0.154
2.0	1.000	0.000	0.999	0.001	0.911	0.089
5.0	1.000	0.000	1.000	0.000	0.961	0.039

excess of, say, 0.1 MeV, e.g. fission spectrum neutrons. RBE values obtained from such experiments will thus apply to the genuine neutron component. If the human body is exposed to such neutrons, a substantially lower RBE will result for the inclusive neutron dose, D', because the exposure is partly due to photons. On the other hand, it needs to be noted in this context that in epidemiological studies, such as follow-up of the atomic bomb survivors, the 'neutron dose' is specified in terms of the 'genuine neutron dose', i.e. only the dose contribution from the protons and heavier charged particles (effectively the neutron tissue kerma within the organ) is counted as the 'neutron dose'.

(249) Let F_n be the fraction of the organ-weighted inclusive neutron-absorbed dose that is due to the genuine neutron dose:

$$F_n = D'_n/D' = \Sigma_T \; w_T D_{T,n}/\Sigma_T \; w_T D_T \tag{4.5}$$

(250) Values of F_n for an anthropomorphic phantom (analogous to those in Table 4.1 for anterior–posterior exposure) are given in Fig. 4.2 for monodirectional anterior-posterior exposure, for exposure with planar-rational symmetry, and for isotropic exposure. They are presented, as are subsequent data, for neutrons between 0.001 and 20 MeV, which includes the important energy range of fission neutrons. F_n decreases rapidly with decreasing neutron energy. In typical moderated fission neutron spectra, neutrons below 1 MeV contribute a major fraction of the dose. The γ-ray component is, therefore, a substantial part of the inclusive neutron dose.

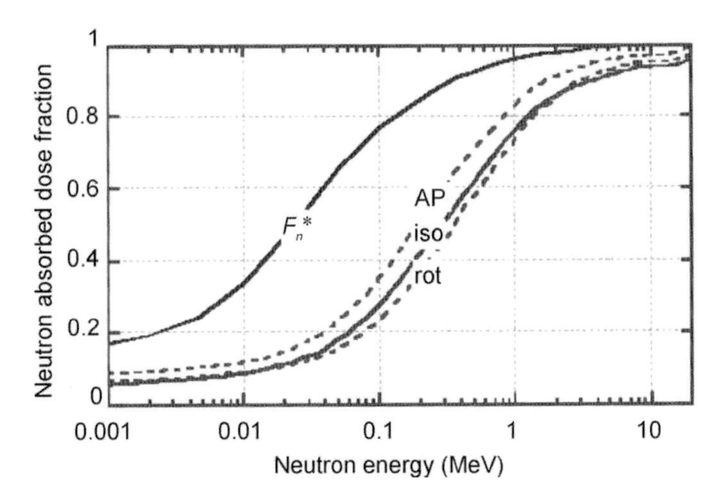

Fig. 4.2. The fractional contribution, F_n, of the genuine neutron dose (neutron kerma) to the effective absorbed dose due to an external field of mono-energetic neutrons of energy E_n. AP, anterior–posterior exposure; iso, isotropic exposure; rot, planar–rotational exposure The upper solid curve represents the analogous parameter for the ambient dose equivalent, H*(10); it shows that the contribution from secondary photons is much smaller with ambient dose than with effective dose (Leuthold et al., 1992, 1997; Mares, 2001, private communication).

4.3.2. The origin of the choice of the radiation weighting factor for neutrons

(251) *Publication 60* (ICRP, 1991) does not comment explicitly on the derivation of the numerical values of w_R, but it states that the w_R values for neutrons are consistent with q^*, the mean quality factor at the reference depth 10 mm of the operational quantity ambient dose equivalent H^*. It also suggests that the value q^* can replace w_R for radiations with no specified w_R.

(252) The w_R for neutrons has been introduced as a step function, with a continuous function offered as 'approximation' (ICRP, 1991). In practice, the continuous function is preferred, and only this continuous dependence is considered here for w_R [see Eq(1.5)]. As shown in Fig. 4.3, the numerical values of w_R are close to q^*. The comparison of the values confirms the conclusion that w_R was chosen to agree essentially with q^*. The actual difference of the values partly reflects a difference between q^* values that were available at the time and values from subsequent computations (Leuthold et al., 1992), and partly reflects the ICRP's choice of a simple numerical approximation.

(253) At the shallow depth of 1 cm in the ICRU sphere, the fractional dose contribution from neutron recoils is, as seen in Fig. 4.2, substantially larger than in the average organ; only between 5 and 20 MeV are the two parameters roughly equal. Accordingly, q^* and thus the current w_R are considerably larger than the effective quality factor, q_E, for whole-body neutron exposures which is defined as the ratio of the effective dose equivalent, H_E, divided by the organ-weighted absorbed dose, D' [see Eq(4.4)]:

$$q_E = H_E / D' \tag{4.6}$$

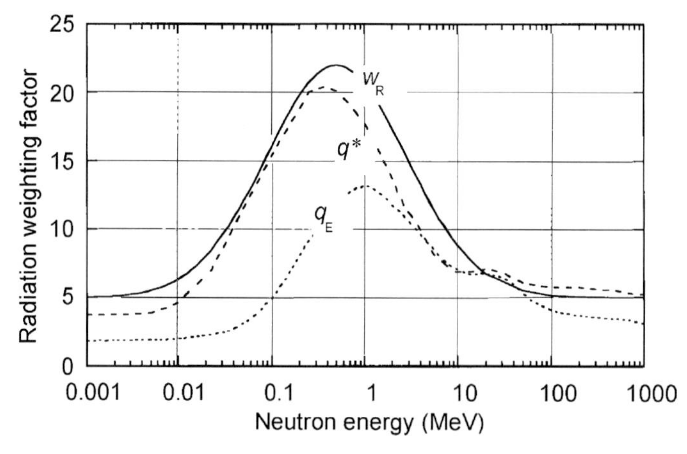

Fig. 4.3. Radiation weighting factor w_R (solid curve) and the ambient quality factor q^* (broken curve). The dotted curve gives the effective quality factor, i.e. the external weighting factor that would have made, for isotropic exposure and with the current w_T and $Q(L)$ values, the effective dose E equal to the quantity effective dose equivalent H_E. [Data for q^* from Leuthold et al. (1992) and for q_E from Mares et al. (1997) for an anthropomorphic phantom for energies beyond 20 MeV interpolated to the values derived by Pelliccioni (1998)].

(254) *Publication 60* (ICRP, 1991) introduced the radiation weighting factor, w_R, as a simplification to avoid unnecessary computations in terms of $Q(L)$. The reason was not that LET was deemed to be an unsuitable biophysical parameter of radiation quality. $Q(L)$ was, accordingly, retained for measurement purposes, and new numerical values were recommended both for w_R and $Q(L)$. Coherence would have been achieved by w_R equal to q_E rather than q^* because both E and q_E relate to the human body or an appropriate phantom of the body. The choice of q_E would have made E nearly equal to the former reference quantity H_E.[8] The difference would merely have been that w_R is independent of the directional distribution of the radiation, while q_E is dependent on it. The dotted line in Fig. 4.3 represents q_E for isotropic exposure. The values for anterior–posterior exposure are somewhat larger, and those for rotational symmetry are somewhat lower. The isotropic case could, therefore, have served as an adequate standard.

(255) It was known that q_E would be less than q^*, but the computed values of q_E for neutrons were not available at the time of *Publication 60* (ICRP, 1991), and this was the reason to pattern w_R after the values of q^*, rather than those of q_E. Setting w_R equal to q^* meant taking insufficient account of the large photon component that results in the human body when it is exposed to neutrons below 1 MeV. The consequence has been a substantial increase of the values of the effective dose from neutrons and a conspicuous exaggeration of w_R values below 1 MeV (Fig. 4.3). The other consequence is that $Q(L)$ fails to serve as an LET dependent weighting factor that is equivalent to w_R. Applying $Q(L)$ to the radiation field in the body leads to a value of the effective dose equivalent that is considerably lower than the effective dose obtained with w_R.

4.4. Options for a modified convention

(256) Three options for a modified convention can be considered:

- a radical simplification of w_R to only two or three numerical values;
- the modification that makes w_R coherent with $Q(L)$ but causes a substantial reduction of the magnitude of the effective dose from neutrons;
- a modification that links w_R to an LET-dependent internal weighting factor without a substantial reduction of the magnitude of the effective dose from neutrons.

4.4.1. Radical simplification of w_R

(257) The system of radiation protection quantities and the numerical conventions that are part of the system must be stable, and unnecessary changes are to be avoided. The need for simplification can, nevertheless, arise whenever the system becomes inflexible and too complicated. One radical simplification that has been

[8] As stated in Section 4.1.2, H_E is taken to be defined with the current parameters $Q(L)$ and w_T.

variously proposed is the reduction of the convention for w_R to just two or three numerical values. Thus, the NRPB (1997) has suggested that a w_R value of 1 be attributed to all photons, electrons, and fast protons, and a value of 10 be attributed to protons and heavier particles. Such a simplification is attractive but there are arguments against it.

(258) The first argument is that it makes little sense to simplify one single aspect in an otherwise complex system. Little would be gained by a radical simplification of the numerical values of w_R unless quantitative risk estimates and precise dosimetry were abandoned in the practice of radiation protection.

(259) The radical simplification of w_R seems impracticable for the added reason that it would tend to force a tightening of dose limits in general. If the current w_R of about 20 for fission neutrons were reduced to 10, this would decrease the numerical value of the effective dose from exposure to fission neutrons by a factor of 2. This would amount to a relaxation of the limits for neutron exposures, which may meet strong objections and would almost certainly generate pressure to offset the change by a decrease of the effective dose limits, which would then apply to all radiations, including photons.

(260) Finally, if the simplified numerical values of w_R continued to be used as external weighting factors, none of the conceptual problems which have been discussed in Section 4.1 would be resolved. This reiterates the fact that the radical simplification would only make sense if it could be part of a generally simplified system of radiation protection with no requirement for precise quantification of dose quantities.

4.4.2. Modification of w_R to establish coherence with $Q(L)$

(261) As stated in Section 4.3.2, E would be largely coherent with H_E and its weighting factor $Q(L)$ if w_R had been chosen to equal q_E rather than q^*. For isotropic exposures to neutrons, if this had been chosen as standard, the values of the two quantities would then be the same. With the current choice, this equality has not been achieved. For an isotropic exposure to 1 MeV neutrons, E exceeds H_E by a factor of 1.6, and for 0.5 MeV neutrons, by a factor of 1.9. For lower neutron energies, the differences are even larger and the high values of E at these neutron energies are clearly in conflict with accepted radiobiological findings.

(262) In view of these considerations, the decision could be taken to adopt the convention that was intended at the outset, i.e. w_R could be set equal to q_E. The dualism of two insufficiently coherent concepts $Q(L)$ and w_R would thus be removed. The external w_R could be used in most practical applications, and $Q(L)$ could be invoked whenever precise determinations, or possibly measurements in a phantom, are required.

(263) Attractive as this procedure might be, it is uncertain whether it is still a viable option. The problem is that it would substantially decrease the current values of the effective dose in the important case of the exposure to fission spectrum neutrons. The current convention has been implemented in practice and has become part of the radiation protection legislation in various nations and also in the European Community. As argued with regard to the radical simplification of w_R, it will

be difficult to justify a major relaxation of the present regulations with regard to occupational exposure to fission neutrons.

(264) In addition, it has been concluded in an analysis (Kellerer and Walsh, 2001) that combined RBE values for fission neutrons from experiments in rodents with the risk data from follow-up of the atomic bomb survivors (see Section 2.3.3) and which accounted for the neutron-induced photon component in the human body, that w_R values for neutrons agree well with the nominal risk coefficient specified by the ICRP (1991). A reduction of w_R would remove this agreement.

(265) These considerations and the need for stability of the radiation protection regulations are judged to outweigh the attractiveness of a modification which would have been the suitable choice in 1990, if the values of q_E had been available then and coherence had been sought with the $Q(L)$ relation adopted (ICRP, 1991).

4.4.3. The proposed modification a moderate numerical change of w_R

(266) As has been pointed out, neutrons induce a substantial photon component in the human body at energies below 1 MeV. Thus at 50 keV, the secondary photons contribute about 80% to the absorbed dose. Since this dose contribution ought to be given the same weight as that from external photons, it follows that the absorbed dose contribution from the high-LET particles is actually weighted by a factor nearly five times larger than the overall w_R value of about 12 seems to indicate at this energy. Evidently an implied weighting factor of 60 or more for the genuine neutron component makes little sense when the corresponding value at 1 MeV is less than 30. While the 50 keV neutrons may have limited practical importance, it needs to be noted that the implied weighting factor is about 50 even at a neutron energy of 200 keV.

(267) The implied magnitude of w_R for neutrons may not be widely recognised but, as pointed out in the preceding section, it has been part of the ICRP system and has been incorporated into the national regulations in those countries that have implemented the current ICRP recommendations. Also, as has been pointed out, the implied higher weighting factor is, apart from the spuriously high values at neutron energies below 0.2 MeV, not in conflict with radiobiological findings. As stated, it may thus be difficult to justify a departure from the current convention that would allow nearly twice the current magnitude of occupational exposures to neutrons in the important energy range between 0.2 and 2 MeV.

(268) It is, thus, advisable to avoid a major reduction of the current values of the effective dose in the energy range of fission neutrons. However, this must not mean to preserve the current values of w_R precisely. The values of w_R below about 0.5 MeV are conspicuously too large compared with the values at higher neutron energies. As has been explained, this is due to the fact that w_R was set equal to q^* (see Fig. 4.3) which does not account adequately for the large dose contribution from secondary photons at low energies of the incident neutrons. The aim must, therefore, be to correct this major inconsistency, but to preserve the current values of the effective dose at energies of the incident neutrons around 1 MeV. This is the energy range of the maximum biological effectiveness of the neutrons and also the neutron energy range of predominant pragmatic importance for occupational radiation protection.

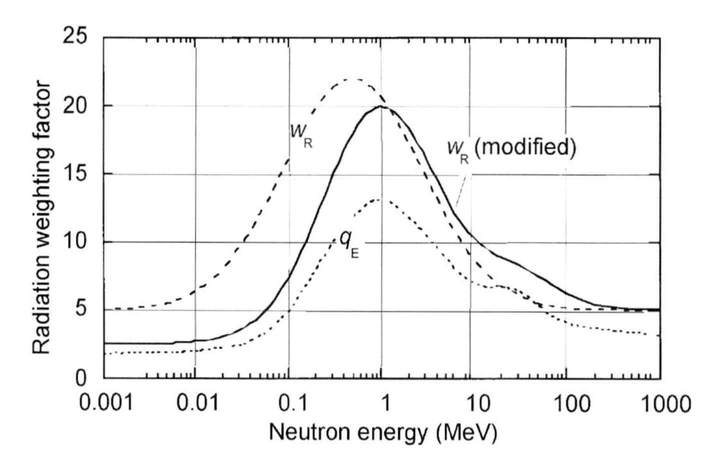

Fig. 4.4. The current radiation weighting factor for mono-energetic neutrons (upper broken curve) and the proposed modification (solid curve). The lower broken curve gives the effective quality factor.

(269) Figure 4.4 presents the proposed modified numerical convention for w_R. It preserves the w_R value at 1 MeV. By being substantially smaller at lower neutron energies, it accounts for the large dose contribution from secondary photons at low neutron energies. At energies between 1 and 100 MeV, the proposed w_R values exceed the current values somewhat. The proposed dependence is in line with the dependence of q_E on neutron energy. But since the current w_R value at neutron energy 1 MeV exceeds q_E by a factor of 1.6, the proposed dependence needs to correspond for all neutron energies to q_E scaled up by a factor of 1.6.

(270) The scaling of q_E cannot be a simple multiplication because this would not preserve the values equal to 1 or close to 1. The proportional increase must, instead, apply to the excess of q_E over 1:

$$w_R = 1 + 1.6(q_E - 1) = 1.6q_E - 0.6 \quad (4.7)$$

(271) The relationship implies that the proposed modified w_R corresponds to the LET-dependent weighting factor 1.6 $Q(L)$–0.6. The numerical equivalence is not precise, since the proposed relation smoothes the dependence q_E somewhat at neutron energies between 20 and 50 MeV. Since w_R is meant to provide simplification, the approximation to q_E is adequate, especially since an uncertain interpolation is still required between the data up to 20 MeV by Mares et al. (1997) and the data above 50 MeV by Pelliccioni (1998).[9]

(272) Due to the broad energy spectra encountered in conventional radiation protection, the overall change in E from neutrons will be modest with the modified convention for w_R except for highly moderated neutron spectra. The required stability

[9] The solid curve in Fig. 4.4 is based on published values of q_E and on Eq. (4.7). Since more computed values of q_E at high energies will emerge, there may be a need for further quantitative comparisons. For computational convenience, the dependence in Fig. 4.4 on neutron energy E (in MeV) can then be expressed as:
$w_R = 2.5[2 - \exp(-4E) + 6\exp(-\ln(E)^2/4) + \exp(-\ln(E/30)^2/2)]$

of radiation protection regulations is, thus, preserved. The essential point is that the substantial decrease at low neutron energies and the slight increase at high neutron energies are consistent with microdosimetry and the available radiobiological information and that the modified dependence of w_R on neutron energy represents a coherent dependence on the underlying biophysical parameter, LET.

4.4.4. The special case of high altitude and space radiation

(273) Neutrons of very high energy are of particular interest with regard to exposures at aviation altitude and in space. There are two distinct peaks in the neutron fluence at about 1.5 and 100 MeV. As shown in the lower panel of Fig. 4.5, the peak around 100 MeV dominates and it contributes most of the neutron-absorbed dose. The neutrons are produced as the high-energy, cosmic, heavy-charged particles—mostly protons and helium ions—penetrate the atmosphere.

(274) Figure 4.6 shows that the maximum build-up of the fluence rate of secondaries is reached at an altitude in the atmosphere of about 20 km, and that at lower altitudes, the neutron, photon, and electron fluences decrease in similar fashion. These secondaries are the predominant contributors to the absorbed dose. The contribution of muons and charged pions is minor at aviation altitudes. The contribution of the heavier charged particles, including their fragmentation products, is likely to amount to about 1% of the effective dose (see Section 3.3 in O'Sullivan, 1999).

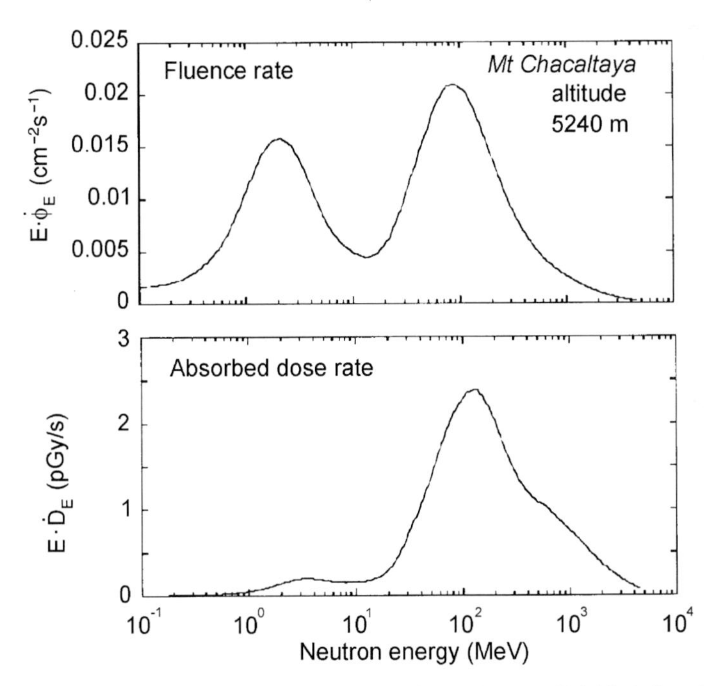

Fig. 4.5. Neutrons at altitude 5 km (Mt. Chacaltaya, Bolivia, 14 GV cut-off rigidity) (based on Fig. 3.2.1 in O'Sullivan, 1999). Upper panel: fluence rate per log-interval of energy. Lower panel: absorbed dose rate to bone marrow per log-interval (fluence to dose conversion factors from Bozkurt et al., 2000, 2001).

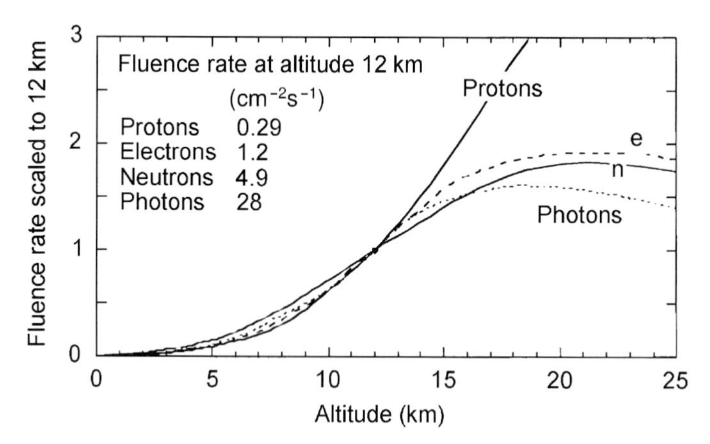

Fig.4.6. The fluence rates of major types of particles vs altitude in the atmosphere for the condition of no geomagnetic shielding and minimum solar activity (based on data in Fig. 3.4.2 in O'Sullivan, 1999).

(275) Photons and electrons pose no problems with regard to the weighting factors. Neutrons are to be assigned the proposed w_R values that are represented in Fig. 4.4; an overall value of $w_R = 6$ for cosmic high-energy neutrons is adequate.

(276) Among the primary cosmic particles, only protons contribute substantially to the absorbed dose at aviation altitudes. With their current $w_R = 5$, they can contribute up to half the effective dose. The protons have received particularly critical attention because different weighting factors have been employed for them by various organisations, and also because the dose contribution from protons varies most strongly with flight altitude and can, thus, be influenced by operational decisions.

(277) In Section 3.4, it was concluded that the current $w_R = 5$ for protons is too high. The conclusion was based on the mean LET of the protons and on radiobiological data for protons up to about 150 MeV. At the markedly higher energies, with a peak around 1 GeV, of the cosmic protons, there is the additional aspect that secondary particles from nuclear interactions need to be taken into consideration. This implies that q_E can exceed the mean quality factor of the protons. Pelliccioni (1998) computed q_E as a function of the incident proton energy. His results are given in Fig. 4.7. It is seen that, according to these calculations, the value at 1 GeV is 1.6 and that this can be taken as a standard value for the cosmic protons. In line with $w_R = 2$ is, therefore, proposed for cosmic proton radiation.

(278) Computations of effective dose in terms of a realistic phantom have been performed and have been used here to infer the corresponding values of w_R. However, the concept of w_R can be unnecessarily complex in high-energy fields. A simpler and more direct approach, including measurements with tissue-equivalent proportional counters (O'Sullivan, 1999), can be employed instead. The ambient radiation field is the equilibrium result of the high-energy cascades generated as the incident cosmic radiation enters the atmosphere. At 12 km altitude, the radiation has already penetrated about 270 g/cm^2; the further degradation in surrounding structures and in the body is only a minor addition that does not substantially alter the spectrum of particles and energies. It is, therefore, possible to compute the

equivalent dose from the radiation free in air without following the degradation steps in the body in detail. When a heavy charged particle contributes to the dose in an organ, it is immaterial whether it has been incident to the body or whether it is due to an interaction within the body. There is, thus, no need to compute the organ-absorbed dose components that correspond to the two cases separately and to weight them differently, i.e. by the w_R for ions in the first case and the w_R for neutrons in the other case.

(279) Figure 4.8 gives the sum distribution of absorbed dose and equivalent dose in LET as the result of particle spectrometry on a transatlantic flight. The low-LET component which contributes the major part of the absorbed dose is not included.

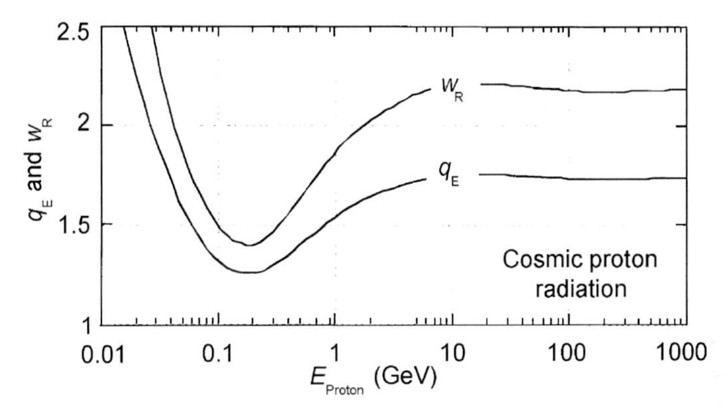

Fig. 4.7. The effective quality factor for high energy protons interpolated from the data by Pelliccioni (1998) and the corresponding radiation weighting factor w_R [see Eq (4.7)]. The proposed standard value is $w_R = 2$.

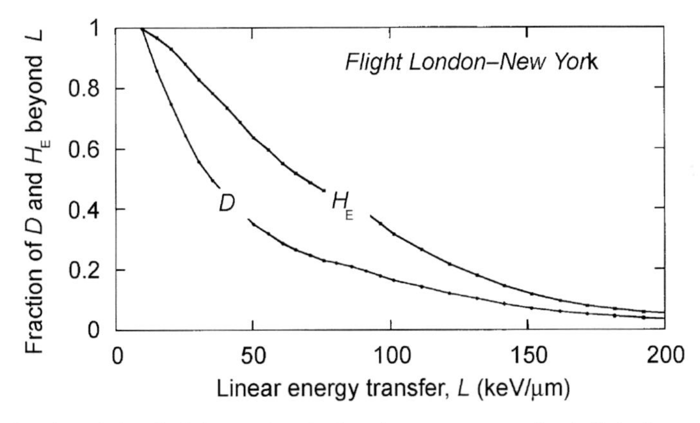

Fig. 4.8. The fraction of absorbed dose and equivalent dose on a transantlantic flight due to particles with linear energy transfer beyond the specified value (based on Fig. 3.2.10 in O'Sullivan, 1999). Only the contribution above $L = 10$ keV/μm is taken into account. The low-linear-energy-transfer (LET) component is excluded; it contributes the major part of the absorbed dose.

(280) Only a minor part of the effective dose due to the high-LET component belongs to the LET region beyond the peak efficiency (≈100 keV/μm). There is no significant dose contribution beyond 200 keV/μm and, thus, no dose contribution from very high-LET particles with greatly decreased biological effectiveness. The somewhat tentatively chosen dependence of $Q(L)$ for L in excess of 100 keV/μm (see Fig. 1.1) is, thus, not highly critical with regard to exposures in aviation. It is, furthermore, apparent from Fig. 1.1 that the LET values lie in a range that can be readily assessed by measurements with tissue-equivalent proportional counters.

(281) The considerations for exposures in space missions must address added complexities because there is, unlike the situation within the atmosphere, no radiation equilibrium. Also, the magnitudes of the exposures can be critical and a more precise assessment will, therefore, be required (NCRP, 2000). There is also a sizable contribution from charged particles heavier than the protons. Their effectiveness is, in view of the still incomplete radiobiological information, difficult to assess.

4.4.5. The continuous and the discontinuous convention

(282) The current w_R has been specified as a step function in neutron energy, and advice has been given that the recommended continuous relationship should be treated as an approximation. This definition has a somewhat awkward consequence. Calculations of organ-equivalent doses or effective dose are usually performed by Monte-Carlo calculations, and the basic input data (e.g. neutron scattering and reaction cross-sections, angular distributions of secondary particles, stopping powers etc.) are continuous in neutron energy, i.e. they never produce discontinuities. Using the step function for w_R introduces artificial steps that are out of line with the remainder of the computations and result in difficulties in practice. For this reason, in all published calculations of E or H_T (ICRP, 1996), the continuous functions had to be accompanied by the disclaimer that they are based on an approximation. Uncertainties in the underlying radiological information should not be a reason to make the definition of a radiation protection quantity impractical. It is, accordingly, advisable to adopt the continuous dependence as the basic convention and to permit the use of the step function as an approximation.

4.4.6. The role of the operational quantities

(283) Since publication of the current ICRP recommendations and introduction of the basic reference quantity effective dose, there have been critical discussions on the relationship between the effective dose, E, and the so-called operational quantities, ambient dose equivalent, $H^*(d)$, and personal dose equivalent, $H_p(d)$.

(284) The operational quantities $H^*(d)$ and $H_p(d)$ have been introduced for dose monitoring in cases of external exposure, either for area monitoring or individual monitoring. Both quantities are phantom-related and are aimed to provide an estimate of E (formerly effective dose equivalent) that is sufficiently conservative in routine situations. Direct measurements of E have never been practicable. In cases where E needs to be assessed with some precision, detailed information about the

radiation field parameters and the exposure geometry must be obtained either to determine a better approximation to E or to quantify the relationship between the operational quantities and E with best accuracy.

(285) For strongly penetrating radiation (e.g. photons above 12 keV or neutrons), the quantity ambient dose equivalent is defined as the dose equivalent at 10 mm depth in the ICRU sphere (under the abstract condition of the 'extended and aligned field'). For photons, $H^*(10)$ always provides a conservative estimate for the effective dose a person would receive if positioned at the reference point. For neutron fields, $H^*(10)$ is conservative between 50 keV and 2.5 MeV, but outside this energy range, E can exceed $H^*(10)$ (ICRP, 1996). At the lower energies, this is an artifact of the current inflated w_R values that will essentially disappear with the modified convention for w_R proposed in the preceding section. At higher neutron energies, $H^*(10)$ can be non-conservative, but for the typical broad neutron spectra at work places, the operational quantity will nearly always be conservative even with anterior-posterior exposure.

(286) A better approximation for neutrons could be achieved if an operational quantity were defined which is more closely related to an anthropomorphic phantom. Such a definition does not exist at present, but the modified convention for w_R and E might facilitate it. However, comparatively simple operational quantities will continue to be employed for monitoring and controlling E and skin dose limits or other defined secondary limits.

5. DETERMINISTIC EFFECTS

5.1. Introduction

(287) Normal tissues exposed to radiation develop effects that are clinically detectable when the dose exceeds what is known as a threshold. For radiation protection purposes, these effects have been designated as deterministic. Both the probability and the severity of deterministic effects increase with dose. For most situations, deterministic effects are prevented by the limits for stochastic effects. It is reasonably assumed that:

> 'the restrictions on effective dose are sufficient to ensure the avoidance of deterministic effects in all body tissues and organs, except the lens of the eye which makes a negligible contribution to the effective dose, and the skin which may well be subject to localised exposures. Separate dose limits are needed for these tissues.' (ICRP, 1991)

(288) In clinical practice, such as radiotherapy, normal tissue effects are of extreme importance and therapeutic regimens must take them into account. Limits for low-linear-energy-transfer (LET) radiation can be based on data for tissue effects obtained from the study of radiotherapy patients, atomic bomb survivors, and, to a lesser extent, clinicians and technicians involved in the early work with radiation sources. The deterministic effects of high-LET radiation in humans are less well documented, but the data available for estimation of relative biological effectiveness (RBE) values have been reviewed extensively (ICRP, 1984; UNSCEAR, 1988; Engels and Wambersie, 1998; Edwards, 1999; IARC, 2000, 2001). Due to the inadequacy of the data available for humans, not only for exposure to neutrons but also to protons and heavy ions, data from animal experiments have to be used in the estimation of some RBE values.

(289) In *Publication 60* (ICRP, 1991), the limits for the lens of the eye and the skin were given in terms of equivalent doses, which by definition involve the radiation weighting factor (w_R) and which are expressed in sieverts. The w_R values are independent of the organ or tissue and of the stochastic effects considered, and are applicable only to conditions of exposure relevant to routine radiation protection and not at dose levels at which deterministic effects occur. ICRP (1991) states:

> 'Equivalent dose is not always the appropriate quantity for use in relation to deterministic effects, because the values of radiation weighting factors have been chosen to reflect the relative biological effectiveness of the different types and energies of radiation in producing stochastic effects. For radiations with a radiation weighting factor larger than 1, the values of RBE for deterministic effects are smaller than those for stochastic effects. The use of equivalent dose to predict deterministic effects for high-LET radiations, e.g. neutrons, will thus lead to overestimates.'

(290) The recommendations of how to adjust doses to take account of radiation quality with regard to deterministic effects are not dealt with in any detail in *Publication 60* (ICRP, 1991), but will be discussed later in this section.

(291) The aim of setting limits for deterministic effects is to prevent them. However, effects at threshold doses evolve from the accumulation of radiation-induced damage at lower doses, and therefore the threshold is really the threshold of detection. The ability to detect damage depends on the methods applied. Cataract is a good example of this problem. Small opacities, i.e. small changes of the lens protein that appear as tiny specs without reduction of visual acuity, are detectable with special methods. In studies on mice, they have been shown to be induced by extremely small doses of high-LET radiation (Bateman et al., 1972; Di Paola et al., 1980), and it appears that they evolve from damage to individual cells that underwent abnormal differentiation. Small opacifications must thus be considered as a stochastic response, while their accumulation, and possibly their interaction, causes the deterministic effect that is noted at an opacification level where visual acuity begins to be impaired. Similar considerations apply to the other deterministic effects that are partly or fully due to cell killing; the killing of individual cells is a stochastic response, but the accumulation of cell killing can lead to deterministic effects.

(292) Depending on the tissue, its cell kinetics, capability for repair and recovery, function, and different critical levels of loss of cells and function are required to reach a clinically detectable effect. In the case of skin, the effect is visible. In the case of other cell-renewal systems, such as gut and bone marrow, the effect is detected by signs and symptoms characteristic of the specific organ. It is assumed that the differences in the effects of different radiation qualities are quantitative and not qualitative, and the evidence bears this out. The RBE for the observed effect reflects, accordingly, the RBE for the underlying cell damage that depends on dose, dose rate, fractionation, and specific cell type as well as the radiation quality. As has been seen with stochastic effects, these factors influence RBE mainly because of their influence on the response to the low-LET reference radiation.

(293) Deterministic effects or normal tissue effects are divided into early and late. The dominant cause of early deterministic effects is radiation-induced loss of cells. Late deterministic effects are also largely due, either directly or indirectly, to cell killing, but, as noted, cataract is an exception.

(294) Variations among tissues in the radiosensitivity of cells, as measured by cell killing, have been considered to be relatively small, although D_o, i.e. the reciprocal initial slope of the survival curve for clonogenic cells and threshold doses for clinical effects vary significantly. The variations of radiosensitivity are, of course, very marked in certain inherited conditions, such as ataxia telangiectasia.

(295) While the ICRP states in the definition of a deterministic effect that such an effect does not occur until a threshold dose is reached, it was, nevertheless, suggested in *Publication 58* (ICRP, 1990) that the RBE value for deterministic effects should be derived by extrapolating the RBE for cell killing to below the threshold dose in order to obtain the ratio of the initial slopes of the responses to the reference radiation and the radiation under study. This approach is analogous to the determination of the RBE at minimal doses (RBE_M) for stochastic effects, and to distinguish the

values, the low-dose RBE for cell killing deterministic effects was denoted by RBE_m rather than RBE_M. Values of RBE_m thus obtained were consistently higher than RBE values based on threshold doses. This is in line with the fact that RBE decreases with dose. The magnitude of the overestimation of RBE will be considered in the next section.

(296) The approach recommended in *Publication 58* (ICRP, 1990) assumes that cell killing is the sole mechanism of induction of deterministic effects. This is obviously not applicable to cataract induction. For other deterministic effects, it disregards the possibility that cell dysfunction or recovery may influence the probability of a clinical effect which may also depend on radiation quality. Deterministic effects, such as erythema, cataract etc., are conventionally described in clinical terms and it is, thus, apparent that the RBE value used to modify the absorbed dose with regard to deterministic effects should ideally be related to the clinical threshold dose for the specific endpoint. The use of RBE_m is, as will be discussed in the subsequent section, a conservative approach with regard to those deterministic effects that are predominantly due to cell loss.

(297) Irrespective of the various complexities and of the method by which RBE is obtained, it is suggested that the weighted dose should be expressed in gray-equivalents (Gy-Eq) with regard to deterministic effects. The dose limits for deterministic effects should likewise be expressed in the weighted dose and in Gy-Eq. Under usual conditions in radiation protection, it will, of course, be sufficient to express the special limits for the skin and the lens of the eye in equivalent dose and in sieverts, which makes no difference for low-LET radiation and is conservative with regard to high-LET radiation. However, in those exceptional radiation protection situations where high-LET effects on the skin or the lens of the eye can be critical, the use of the weighted dose in Gy-Eq is appropriate.

5.2. ICRP *Publication 58*

5.2.1. Aim of the report

(298) In 1987, an ICRP Task Group was appointed to examine the question of RBE for deterministic effects, and their report was published as *Publication 58* (ICRP, 1990). The Task Group stated the purpose of its report as follows:

> 'In view of the specific purpose for which Q values were selected, it is evident that they do not represent the highest values of RBE judged to be applicable to all effects in all tissues. Thus, for some of the many types of effects, either stochastic or deterministic, RBE values for specific exposure conditions might be larger than Q values for the high-LET radiation considered. For a given tissue exposed selectively, e.g. as a consequence of the intake of a radionuclide specifically retained in this tissue, dose limits or limits in intake based on Q values might not be adequate to prevent deterministic effects induced by high-LET radiation if an effect in this tissue is caused with a very high RBE value for these conditions of exposure. It is therefore of interest to analyse the data in the

literature on RBE for deterministic effects in individual tissues to judge whether for chronic irradiations or in accidental exposures, specific high RBE values are appropriate for the estimation of possible health consequences. In the present report, data on RBE values for effects in tissues of experimental animals and man will be analysed to assess whether for specific tissues the present dose limits or annual limits of intake based on Q values are adequate to prevent deterministic effects.'

(299) The report contains a very useful review of deterministic effects, which draws extensively from the literature on the effects commonly referred to as normal tissue effects. This designation arose because of the interest and importance to radiotherapists of the differences in the responses of cancerous and normal tissue.

(300) To determine a weighting factor for radiation quality with regard to deterministic effects is, in one respect, simpler than the determination of a weighting factor for stochastic effects. RBE values can be obtained at doses of the threshold level for individual deterministic effects, which means that no low-dose extrapolation is needed, and that RBE values are generally lower than those for stochastic effects. On the other hand, the task is more complex because there are more different RBE values in different tissues for different endpoints. There is also the difficulty that threshold doses vary between individuals and are not always easily determined.

(301) To simplify the issue, *Publication 58* (ICRP, 1990) recommended reference to the low dose limit of RBE even for deterministic effects, although this entailed extrapolation to doses at which the responses to both the radiation under study and the reference radiation were below the threshold. To make this type of extrapolation possible, it was assumed that all relevant deterministic effects depend on cell killing and that, accordingly, the RBE for deterministic effects can be related to cell killing, regardless of the specified endpoint. To distinguish the maximum values of RBE for the two categories of radiation effects, the notation RBE_m was used for deterministic effects.

5.2.2. Linkage of RBE_m to cell killing

(302) *Publication 58* (ICRP, 1990) derived RBE values for deterministic effects under the assumption that these effects are due to cell killing. It postulated that the magnitude, $E(D)$, of a deterministic effect at dose D depends on cell survival, S, at this dose, and only through cell survival on radiation quality. This means that the functional dependence, $E'(S)$, of the effect level on cell survival, S, is independent of radiation quality:

$$E(D) = E'(S) \tag{5.1}$$

(303) Since $S(D)$ determines the effect, it is called a metameter. Equal values of the metameter imply equal effect regardless of the numerical form of the dependence of the observed effect on cell survival.

(304) Cell survival can be expressed in terms of a linear-quadratic response in the dose, D_L, of the reference low-LET radiation or the dose, D_H, of the high-LET radiation:

$$S_L(D_L) = \exp\left(-\left(\alpha_L D_L + \beta_L D_L^2\right)\right) \text{ and } S_H(D_H) = \exp\left(-\left(\alpha_H D_H + \beta_H D_H^2\right)\right) \tag{5.2}$$

The assumption of this model is that loss of reproductive capacity can be caused by damage from a single track or by an accumulation of damage caused by two or more particle tracks. In the case of the high-LET radiations with exponential survival curves, the value of β_H is decreased, and at very high LETs, it can be considered to be negligible. The equal-effect condition is then:

$$\alpha_L D_L + \beta_L D_L^2 = \alpha_H D_H \quad \text{or}: \quad D_H = \alpha_L/\alpha_H D_L(1 + D_L/(\alpha_L/\beta_L)) \tag{5.3}$$

RBE is, thus:

$$\text{RBE} = D_L/D_H = \alpha_H/\alpha_L/(1 + D_L/\theta) \text{ with}: \quad \theta = \alpha_L/\beta_L \tag{5.4}$$

and the low dose limit is $\text{RBE}_m = \alpha_H/\alpha_L$.

(305) θ is termed the 'crossover dose'. It is equal to the dose at which the linear and dose-squared terms contribute equally to cell inactivation. θ is a important reference parameter for the response of a tissue. The α component determines the initial slope, and the β component determines the contribution by accumulation of damage. Small values of θ indicate, therefore, high recovery capacity, and large values indicate nearly linear response with little recovery.

(306) *Publication 58* (ICRP, 1990) shows how to infer RBE_m in terms of the known or estimated θ from the RBE observed at the dose D_L of the low-LET radiation. If the quadratic component is taken to be the same for the high- and the low-LET radiation, the following equation is obtained:[10]

$$\text{RBE}_m = \text{RBE} \cdot \left[1 + (D_L/\theta)\left(1 - 1/\text{RBE}^2\right)\right] \tag{5.5}$$

(307) If, as is usually assumed, the quadratic component can be disregarded for the high-LET radiation, one obtains from Eq(5.4) the same relationship that has been derived for RBE_M in Chapter 2 [Eq(2.5)]:

$$\text{RBE}_m = \text{RBE}(1 + D_L/\theta) \tag{5.6}$$

(308) In line with the statement in Chapter 2, it has already been noted in *Publication 58* (ICRP, 1990) that the term $(1 + D_L/\theta)$ is analogous to the dose and dose-rate effectiveness factor.

[10] In *Publication 58* (ICRP, 1990), the corresponding equation (3.11) contains a misprint: the term α_L/β_L must be substituted for $\alpha_L\ \beta_L$.

(309) The last term $[1 + D_L/(\alpha_L/\beta_L)]$ has been called the relative effectiveness factor (REF) in applications to normal tissue tolerance in radiotherapy because it represents the increase of the relative effectiveness of a fractionated treatment with doses per fraction, D_L, relative to the effectiveness of infinitely small fractions, applied with long intervals, whereby only the linear term contributes to cell reproductive death (ICRP, 1984). This REF for normal tissue damage is equivalent to the dose-rate reduction factor applied in estimates to derive tumour induction risk at low doses and dose rates from data at large doses.

(310) The extrapolation of the response, i.e. of the metameter, to doses below the threshold of the observed effect is somewhat artificial, but can serve as a convenient simplification. The question is whether the method that involves determining the RBE of a radiation on the tissue or organ level in terms of cell killing will be too conservative, i.e. will result in an inappropriate or unnecessarily stringent dose limit for the densely ionising radiation, if it is based on the estimation of the effectiveness of the two radiation qualities below the threshold levels.

(311) Let D_t be the assumed threshold dose (or a relevant value of the organ dose limit), Eq(5.6) provides the factor of overestimation which happens to equal $\text{REF} = (1 + D_t/\theta)$. *Publication 58* (ICRP, 1990) gives θ values for different tissues that range from 2 to 10 Gy. If $D_t = 0.5$ Gy is assumed, the overestimation factor will vary between 1.25 and 1.05, which suggests that the extrapolation to low doses will not usually increase the RBE_m values unduly. As an example, θ values of 5 and 10 Gy are given for skin, which implies an overestimation by a factor of only 1.2 or 1.1, even if D_t is taken to be 1 Gy.

(312) However, it must be kept in mind that the argument depends on the assumption that the deterministic effects primarily reflect cell killing and its linear-quadratic dose dependence. This assumption is clearly not applicable to the case of cataract formation but, as emphasised in the next section, it needs to be questioned more generally. As a matter of fact, the values of RBE_m derived by the approach used in *Publication 58* (ICRP, 1990) are consistently higher than those based on threshold doses. The average of the listed values of RBE_m for early and late deterministic effects after exposure to 1–5 MeV neutrons is 6.2 and 8.3, respectively, whereas the corresponding average RBE estimates are 4.8 and 5.4. In the case of 5–50 MeV neutrons, the average values of RBE_m for early and late effects are 3.3 and 5.3, respectively, whereas the corresponding average RBE estimates are 2.6 and 3.3.

(313) That the RBE_m values are higher than the experimentally observed values of RBE agrees with expectations. It remains to be explored whether the observed values are obtained at doses close enough to presumptive threshold doses to be more relevant to radiation protection situations than RBE_m.

5.2.3. Possible complexities

(314) The treatment in *Publication 58* (ICRP, 1990) is predicated on the assumption that the underlying radiation-induced lesions below the threshold behave as predicted by the linear-quadratic model and that cell killing is the only cause of deterministic effects. As explained in the preceding section, the approach can be

acceptable under this assumption because the initial slope predominates even in the dose range of the threshold or the organ dose limits, which means that the extrapolation to low doses causes little numerical change.

(315) It must, of course, be realised that the considerations refer to moderately large doses, i.e. to the general magnitude of threshold doses for deterministic effects. It is, therefore, of no concern whether there are insufficiently known complexities at lower doses that lead to deviations from the linear-quadratic dose relationship. While the appropriateness of the linear-quadratic model for the cell-killing curve in its initial low-dose part has recently been questioned (Joiner et al., 1996; Wouters and Skarsgard, 1997), there is no need to modify the value of RBE_m in view of this possible complexity.

(316) The extrapolation below the thresholds of deterministic effects is, accordingly, of minor concern for determination of RBE_m. The major concern is that the linkage of RBE_m to cell killing must be questioned. For the lens of the eye, i.e. for cataract formation, cell killing is clearly not the relevant endpoint and the RBE for cataract induction is, therefore, treated separately in the subsequent section. However, even for other deterministic effects, RBE need not be entirely equal to the value for cell killing. It is the aim of radiation protection to prevent clinically significant deterministic effects that relate more importantly to function than to a numerical change in cell populations. There is more biology in the response of the tissue to a radiation exposure than the proliferative response of the cells, and if the additional factors that influence recovery at the tissue level are differently influenced by high-LET radiation, the RBE for the tissue reaction may differ from that for cell killing alone.

(317) A number of other considerations have to be taken into account. For example, RBE values for late effects are higher than those for acute effects. As noted, the ICRP Task Group related the RBE for deterministic effects entirely to acute cell killing. The higher RBE values for effects occurring a long time after the exposure would, thus, imply that these effects are generally associated with lower doses, but there is little evidence that this is so.

(318) The RBE varies with neutron energy and with the LET of heavy ions. The available data for taking such factors into account are scanty. The results reported in *Publication 58* (ICRP, 1990) illustrate the tissue dependency and the differences between early and late effects. The average RBE_m for 1–5 MeV neutrons is 6.2 (in four tissues) for early effects and 8.3 (in six tissues) for late effects; for 5–50 MeV neutrons, the values are 3.3 and 5.3 for early and late effects, respectively. These results also indicate the inverse relationship of RBE to neutron energy.

(319) There have been other approaches to the calculation of RBE, for example, the so-called biological weighting function (Paganetti et al., 1997) based on a biophysical model and microdosimetric parameters (Zaider and Brenner, 1985; Morstin et al., 1989). In this approach, it is assumed that the dose–effect relationship can be expressed as an integral over two separate functions, one of which describes the distribution of energy in the target, and the other describes the relevant cellular response. Paganetti et al. (1997, 2002) applied this approach to determine the RBE of protons, assuming that the dose–response relationships at low doses are linear.

(320) To define RBE_m simply as the ratio of the threshold doses would have the advantage that no model assumptions have to be made, but it would require a specification and quantitative assessment of the threshold dose. The specification will have to include the reason for selecting the method that is applied to detect the minimal effect. Since different methods identify different minimal effect levels, they will also imply different threshold doses. The reference to the use of the ratio of threshold doses requires a critical examination, especially of the validity of the estimates of threshold doses currently available. While there is information on threshold doses for a number of effects induced by low-LET radiation, there are few data for neutrons. Nevertheless, in spite of these problems, RBE values based on threshold doses appear to be an appropriate reference for radiation protection purposes with regard to single acute doses. In the case of fractionated doses, RBE must be based on the threshold dose for small fractions.

(321) RBE and RBE_m data for early and late deterministic effects in experimental animals after exposure to neutrons of various energies (ICRP, 1990) indicate the following: RBE values are tissue specific, dependent on the neutron energy, and are lowest with single doses and increase with fractionation. The latter is so because the fractionation regimens result, due to the linear-quadratic response, in a lower effect of the low-LET reference radiation. *Publication 58* (ICRP, 1990) listed RBE values based on 'the relative effectiveness in causing some specified endpoint, for example, moist desquamation in skin after either single doses of 10–20 Gy, and of multiple fractions of 2–3 Gy of photons' and of RBE based on 'the ratio of the α coefficients for the responses of the reference radiation and the radiation under study in the linear-quadratic model'.

(322) RBE values for eight tissues in mouse, rat, and pig were given for early and late responses exposed to neutrons of 1–5 MeV mean energy. A second group of values was given for the responses of 11 tissues to neutrons of 5–50 MeV mean energy. Since this group includes the neutron energies used in radiotherapy, some results for humans were given along with those from animal experiments. The average RBE_m values for early and late effects are 4.0 and 5.5, respectively, and the corresponding average RBE estimates are 2.6 and 3.3. Despite the variation in the radiation regimens and in individual estimates, it is clear that the values of RBE_m are higher than the estimates of RBE. A salient question about the experimental techniques is whether sufficiently small fractions have been used to obtain a limiting value of RBE. In a number of tissues, the effects appear to correlate with the effects on the clonogenic cells of the tissue. For example, the estimate of moist desquamation in skin, which is an in-vivo effect appears to correlate with the estimates of survival of the clonogenic cells based on the linear-quadratic model. Such findings confirm that cell killing is the central mechanism of the induction of, at least, some early deterministic effects.

(323) To obtain a limiting value for RBE_m requires exposure to very small fractions, a requirement not often fulfilled. However, there is no need to derive RBE_m if RBE values can be obtained for low-LET dose fractions equal to the relevant annual occupational limit or to the assumed threshold for the deterministic effect. The ICRP (1991) recommended an annual occupational dose limit for the skin of 0.5 Sv.

If this value, i.e. 0.5 Gy, is sufficiently conservative with low-LET radiation, 0.5 Sv is overconservative with regard to high-LET radiation. With regard to skin, it will, thus, be appropriate to use an annual occupational dose limit from densely ionising radiation 0.5 Gy divided by a weighting factor based on RBE values from clinical or experimental experience against the low-LET absorbed dose of 0.5 Gy.

5.3. RBE for lens opacifications and cataracts

(324) ICRP has classified lens opacification as a deterministic effect for radiological protection purposes based on the report of Merriam and Focht (1957) that no cataracts were detected in their study at doses below 2 Gy.

(325) The mechanism of cataract induction by ionising radiation is not completely understood but it involves a lesion in the proliferative cells of the germinative zone of the lens. This lesion causes an abnormal differentiation that results in the abnormal lens fibres which in the early stages can be detected in the posterior subcapsular region, a distinguishing feature of radiation induction. Some experts consider that the early small opacities progress to a size that may reduce visual acuity. The experimental data and some recent studies on humans exposed to low doses also suggest that there may be no threshold, or one so small that it cannot be detected. Such findings raise the question of whether lens opacification should be classified as a deterministic effect.

(326) The recent studies on humans have been reviewed by Shore and Worgul (1999) and indicate that small opacities can be detected after exposure to much smaller doses than 2 Gy. It is not clear whether a threshold or a linear-quadratic response describes the data best. It has long been known that there are a number of modifying factors, such as dose rate. In patients irradiated prior to bone marrow

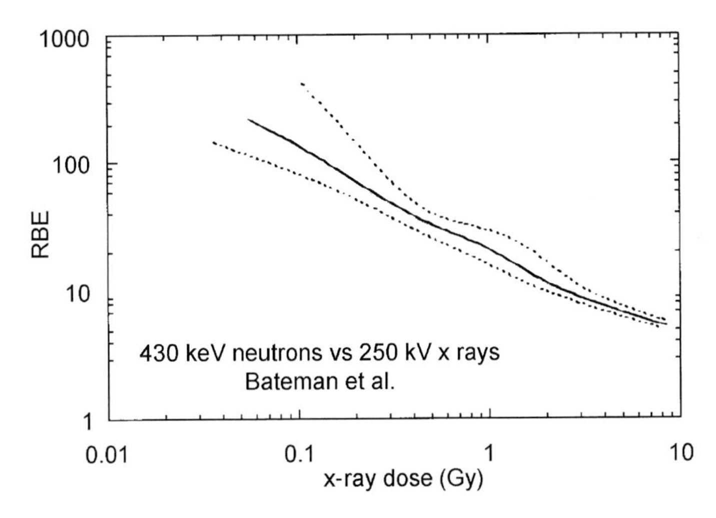

Fig. 5.1. Relative biological effectiveness (RBE) for lens opacification by 430 keV neutrons (Bateman et al., 1972). The diagram is replotted from the RBE vs neutron dose diagram (Kellerer and Rossi, 1982). The estimated values (solid curve) and the standard error range (dotted curves) are indicated.

Table 5.1. Relative biological effectiveness of neutrons relative to 250 kV x rays for lens opacification in mice[a]

Neutron energy (MeV)	RBE against specified dose of reference radiation (250 kV x rays)				Reference
	0.15 Gy	0.3 Gy	1 Gy	2 Gy	
0.430	95 (64–230)	39 (29–68)	21 (15–29)	12 (9.5–16)	Bateman et al. (1972)
1.8	> 20	26 (15–42)	11 (6–18)	7 (3.5–12)	
14	19 (9.5–50)	9.6±3.3	3.5±1.5	–	
0.440	>350	>350	32±11	13±4	Worgul et al. (1996)
1	> 45	45±6	25±1.5	23±1.3	
5	> 25	23±4	18±1.2	14±1.5	
15	> 40	39±7	11±1	9±0.9	Di Paola et al. (1980)
400	–	7±0.8	7±0.4	6±0.4	
600	–	–	7±0.8	6±0.2	

[a] Values of RBE and standard errors interpolated from published data; asymmetrical standard error ranges are given in parentheses.

transplantation, the induction of cataracts was reduced when the dose rate was decreased and when the irradiations were fractionated (Belkacemi et al., 1996). These studies also suggested a greater susceptibility to radiation cataractogenesis in children.

(327) The human data for the effects of radiation qualities other than low-LET radiation are sparse. Some of the physicists involved with the early cyclotrons developed cataracts. It was thought that they were exposed to less than 1 Gy of a mixed neutron and γ-field. Some patients treated with 12 fractions of 7.5 MeV neutron incurred cataracts with some loss of vision (Roth et al., 1976). Compared to the effect of fractionated exposures to x rays that were reported by Merriam and Focht (1962), the RBE for these neutrons appeared to be about 3. Otake and Schull (1990) fitted various models to the responses of the lens in atomic bomb survivors, based on the DS86 dosimetry. The RBE for induction of cataracts by the neutron component was estimated to be 32 (confidence range: 12–89). However, as this was based on a linear no-threshold response for both γ and neutron radiation, the RBE would be higher if a linear-quadratic model was used for the γ radiation.

(328) For protons, the evidence from the studies of Rhesus monkeys is that the cataractogenic effects are similar to those of photons (Niemer-Tucker et al., 1999). For neutrons, the results of experiments with mice (Bateman et al., 1972; Di Paola et al., 1980; Worgul et al., 1996) indicate high values of RBE. For neutrons at their most effective energy of about 400 keV, Bateman et al. (1972) and Worgul et al. (1996) reported RBE values, compared to 0.15 Gy of 250 kV x rays, of about 95 and 350, respectively (see Fig. 5.1 and Table 5.1). It is difficult to assess how these results relate to the RBE for clinically significant lesions in humans.

(329) In the case of heavy ions, there are no data for humans exposed to heavy ions alone. However a recent report of the incidence of cataracts in astronauts (Cucinotta et al., 2001) has raised the possibility of a high RBE for heavy ions or a much higher sensitivity to protons than is suggested by the data for monkeys.

Experimental results indicate high RBEs for heavy ions. Brenner et al. (1993) reported RBEs of 50–200 for both iron (190 keV/μm) and argon (88 keV/μm) ions compared to 250 kV x rays.

5.4. Non-cancer late effects

(330) In considerations on doses relevant to radiation protection, deterministic or non-cancer effects have not been included in the summary risk. The reason is that the threshold doses are believed to be sufficiently high to preclude most of these effects and, moreover, that there has been little indication of their contribution to mortality. With the exception of whole body exposures as a result of an accident, radiation-induced deterministic effects are specific to a tissue or organ. Recently, non-cancer mortality in the atomic bomb survivors was re-assessed (Shimizu et al., 1999), and non-cancer effects were found to have caused excess mortality at weighted doses above 0.5 Sv. Earlier, a threshold of 1 Sv had been estimated for radiation-induced non-cancer mortality (Shimizu et al., 1992). The shape of the dose–response curve cannot be reliably determined, but as the excess mortality is attributed to deterministic effects, a threshold-type response would be expected. With the current radiation protection dose limits, the lifetime occupational effective dose could theoretically reach 1 Sv. If this value could actually be reached in practice, there might be concern about using only stochastic effects as the basis of risk and protection standards for worker populations. On the other hand, there is some reason to assume that the threshold doses would be higher for the highly protracted lifetime exposure. In actual practice, the regulations will ensure that the maximum lifetime effective doses are much lower than the theoretical 1 Sv. Thus, there is presently no need to reconsider limits of the effective dose in view of potentially increased non-cancer mortality. However, the issue continues to deserve attention. The change in the estimation of excess mortality from non-cancer effects since the first report (Shimizu et al., 1992) is sufficiently large to alert one to the trend that will become apparent from the 1990–1997 data (Preston et al., 2003).

6. CONCLUSIONS

6.1. Problems with the concept of RBE

(331) The relative biological effectiveness (RBE) of a densely ionising radiation relative to a low-linear-energy-transfer (LET) reference radiation has remained, for many decades, the concept used for comparing the dose levels of different radiation qualities to cause the same level of effect for a specific endpoint. The assessment of the low-dose maximum RBE, RBE_M, has been central to the selection of quality factor [$Q(L)$] and radiation weighting factor (w_R) values.

(332) Being a ratio, the RBE is subject to changes and uncertainties in both numerator and denominator. Thus RBE can be misleading if its value is seen as a measure of the effectiveness of the densely ionising radiation, i.e. if it is overlooked that its magnitude reflects equally, although in a reciprocal way, the effectiveness of the reference low-LET radiation at low doses that is strongly influenced by a number of factors and often much more difficult to determine. The uncertainty of RBE_M predominantly reflects the influence of these factors and not so much the uncertainty of the effectiveness of the high-LET radiation.

(333) The same problem relates equally to $Q(L)$ and w_R, the two parameters that are intended to represent the magnitude of RBE_M for late stochastic radiation effects, especially cancer, in man. The uncertainty that underlies the selection of these two standards includes the choice of the low-dose extrapolation that is used to obtain the risk estimate of the reference radiation.

6.2. Need to invoke experimental data

(334) Apart from the problem of the correct interpretation of the concept of RBE, the main difficulty from a radiation protection point of view is that direct data from humans are available only for certain α emitters, such as radon daughters, radium, and, more recently, plutonium. For fast neutrons or heavy ions, there are no adequate data from humans, and data from experimental systems must, thus, be utilised. Determination of the maximum value, RBE_M, of the neutron RBE from experimental studies has, therefore, become the principal method for obtaining the dose weighting factor for radiation qualities for which there is no direct information on their stochastic effect in humans. The approach is not ideal, but no alternative method has been identified.

(335) For stochastic effects, it has been seen as essential to use an RBE_M that is difficult to determine with acceptable confidence limits, especially for complex endpoints on the tissue level, such as cancer. The paucity of data that are useable for the determination of RBE_M values makes the selection of w_R and the assignment of the relationship of Q to either unrestricted LET, L, or its microdosimetric analogue lineal energy, y, a daunting task. For radiations such as higher energy neutrons or heavy ions, there is a lack not only of human data, but also a lack of experimental information on cancer induction. Additional information related to cellular endpoints, such as chromosome aberrations, must, thus, be taken into account.

(336) Although not stated in *Publication 60* (ICRP, 1991), it needs to be noted that the selection of w_R has been related to the new convention for $Q(L)$ while $Q(L)$, in turn, has been patterned after the relationship $Q(y)$ which was recommended by the Joint ICRP and ICRU Task Group (ICRU, 1986) in line with their assessment and the earlier evaluation (Sinclair, 1985) of the relevant radiobiological data.

6.3. Two approaches towards the determination of RBE_M

(337) The familiar approach links $Q(L)$ and w_R to a representative value, RBE_M, of the maximum RBE reached at low doses or low dose rates. The values of RBE_M for different radiations can be determined, at least in principle, from experimental data, and there are two ways to do this.

6.3.1. The low-dose method

(338) For the experimental determination of RBE_M values, studies need to be performed at low doses or low dose rates. Such studies are feasible for certain cellular endpoints, such as chromosome aberrations, but low-dose tumour studies in animals are difficult and costly. Major experiments with acute exposures, including low neutron doses, low dose rates, and fractionation, have been performed. Life-shortening experiments are less difficult to perform and they can serve as a proxy for excess cancer mortality studies.

(339) RBE_M values have been estimated in terms of life shortening in mice at dose rates sufficiently low that the dose dependence for both radiations can be assumed to be linear. RBE is then independent of dose and can be taken to equal RBE_M. The advantages of life-shortening studies appear to outweigh their drawbacks. For male mice and long-term protracted exposures, values for fission neutrons against γ rays between 17 and 42 have been obtained for different radiation regimens.

(340) The fact that much of the tumour data in mice has been determined in female mice is noted, and it is suggested that such data should be used with caution. An examination should be made of the effect which inactivation of the exquisitely sensitive mouse ovary exerts on the incidence and multiplicity of various types of tumours.

(341) Experiments in a number of non-cancer systems have been performed with low-dose acute neutron exposures, and RBE values against photon doses of 1 Gy or less have been determined. Fairly good values of RBE_M have been obtained for the induction of dicentric chromosomes; with values of about 70 of the neutron RBE against γ rays. However, the dose relationship for this system exhibits large curvature. The neutron RBE against 1 Gy γ rays is only about 12–15 (see Fig. 3.3). If the low-LET dose response for solid cancers were similar to that for the induction of dicentric chromosome aberrations, its curvature would have to be very marked [dose and dose-rate effectiveness factor (DDREF) = 5]. There is no indication of such curvature in the epidemiological data for the atomic bomb survivors. It is, thus, uncertain whether the low-dose RBE for induction of dicentric chromosome aberrations is relevant to human solid cancer.

6.3.2. The high-dose method

(342) The CIRRPC panel and the NRPB proposed to determine the 'high-dose' value, RBE_H, of the neutron RBE against a γ-ray dose that is sufficiently high to permit the reliable determination of RBE_H, and then to make a judgement—on a broader basis of experimental and epidemiological information—on a standard modifying factor which is termed the DDREF in *Publication 60* (ICRP, 1991). The low dose limit, RBE_M, of the neutron RBE, is then inferred by multiplying DDREF into the observed value, RBE_H, of the 'high-dose' neutron RBE. The NRPB proposed to use a DDREF value of 2 that was chosen in *Publication 60* (ICRP, 1991) to derive their nominal risk coefficient for low-LET radiation. The uncertainty of the value of DDREF has been discussed in NCRP (1997), and it is not clear whether these uncertainties that are inherent in the high-dose method prescribed by CIRRPC and NRPB are less than the uncertainties of the experimental determinations of RBE_M. In order to indicate that it is a value of RBE_M obtained by this specific procedure, a separate notation, RBE_A, has been employed by the NRPB.

(343) In experiments with 430 keV neutrons on the induction of benign mammary tumours in female Sprague-Dawley rats, RBE values of about 50 have been obtained for RBE_H against an x-ray dose of 1 Gy (Shellabarger et al., 1980). However, the relevance of these results is uncertain. Experiments on the induction of non-lethal and lethal tumours in male Sprague-Dawley rats by small acute doses of fission neutrons have provided an RBE_H value of 50 against a γ-ray dose of 1 Gy (Lafuma et al., 1989; Wolf et al., 2000). Evaluation of the same experimental data in terms of life shortening has provided a neutron RBE against 1 Gy γ rays of about 30, which indicates that life shortening in rats may be associated with somewhat lower RBE values than the induction of lethal tumours and of lung tumours which happen to be non-lethal.

6.4. Other uses of RBE

(344) RBE is treated in this report because it provides guidance on the selection of weighting factors for radiation quality. The weighting factors relate to low doses, and the report is, therefore, primarily concerned with RBE values at low doses, i.e. with RBE_M.

(345) There are, however, other uses of RBE. One application concerns the derivation of the risk factor of high-LET radiation. For this purpose, RBE values, e.g. from animal experiments with neutrons, are combined with epidemiological data for γ rays. If this is done in terms of RBE_M, the resulting risk estimate for neutrons is subject to considerable uncertainties, **which primarily reflect the lack of firm data for the low-dose and low-dose-rate effect of γ rays**. In the high-dose method, the γ-ray risk coefficient (risk divided by dose) observed at a high dose is multiplied by the neutron RBE which is observed against this dose in animal experiments; the risk coefficient for neutrons is, thus, obtained. The application of this method in terms of the solid cancer mortality data of the atomic bomb survivors, and RBE information from life shortening in male mice and tumour induction in male rats, has provided a

risk estimate for fission neutrons that agrees well with the ICRP's nominal risk coefficient for solid tumours and with the current w_R value for neutrons (Kellerer and Walsh, 2002).

(346) The use of RBE values with regard to higher doses is also required for calculating probabilities of causation of cancer. In order to avoid confusion, it has been suggested that standard radiation weighting factors for this particular purpose be termed differently. The term 'radiation effectiveness factor' has been suggested (Kocher, 2001). This choice conflicts with an earlier different connotation of the term (ICRP, 1990) which perhaps should be abandoned now.

6.5. Radiation weighting factor and quality factor–conceptual problems

(347) Apart from the issue of how w_R and $Q(L)$ are linked to RBE, questions have arisen that relate, on the one hand, to the reason for introducing the new and separate concept w_R, and, on the other hand, to the relationship between the numerical values of w_R and $Q(L)$.

(348) The replacement of the former reference quantity effective dose equivalent, H_E, by the effective dose, E, has led to problems concerning the inter-relationship of w_R which enters the definition of E and $Q(L)$ which is still required in measurements. These problems and the issue of the relationship between E and the operational quantities H^* and H_p have been the subject of controversy.

6.5.1. Need for rigorous definitions and coherent concepts

(349) When the ICRP linked dose limits or average doses for radiation workers to nominal risk coefficients, it provided a tool for a realistic perception of radiation risks and a useful guideline for the setting of limits and for the comparison of risks from different sources. It was duly emphasised and it was also expressed in the designation of a nominal risk coefficient that there are uncertainties in the extrapolation of observations at high doses to the small and often minute doses that are at issue in the radiation protection of workers or the public. These uncertainties go beyond the imprecision of numbers that is due to imperfect observations and statistical errors; they are, instead, primarily determined by the lack of accuracy that stems from plausible, but unproven, assumptions and extrapolations that are required in any workable and sufficiently conservative system of risk assessment.

(350) If employed with proper judgement, the nominal risk coefficient is a useful tool. If inappropriately applied, it can be used for calculating numbers of fatalities in large populations exposed to trivial doses, and such numbers can further an already distorted perception of radiation risks. The ICRP is responding to this problem by exploring new concepts that help to avoid misuse. Also, to alert one to the danger of overinterpretation, the ICRP has stated that the detail and precision inherent in the formal Q–L relationship is not justified because of the uncertainties in radiological information. However, it would be a misconception to conclude from this statement that lack of rigour is acceptable in the concepts and in the numerical conventions for radiological protection.

(351) The reason is apparent: the current system of radiation protection uses general principles, such as optimisation, that do not require precise quantification, but it uses, in addition, recommended exposure limits that necessitate demonstration of compliance through measurements and computations. Such measurements and computations need to be coherent, and in critical situations, they need to be sufficiently precise to avoid controversy. When concepts are blurred and definitions lack rigour, ambiguities and discrepancies are bound to arise that complicate computations and measurements rather than simplifying them.

6.5.2. Impracticability of a radical simplification

(352) The system of radiation protection tends to be criticised for being unduly complicated. One radical simplification that has been proposed is the reduction of the convention for w_R to just two numerical values; $w_R = 1$ would be attributed to all photons and electrons and to fast protons, while $w_R = 10$ would be attributed to protons and heavier particles. The proposal appears attractive, but there are reasons against it.

(353) First, there needs to be balance, i.e. it makes little sense to simplify one single aspect in an otherwise complex system. To justify the radical simplification of w_R, analogous simplification would need to be provided with regard to dosimetry, risk estimation, and the setting of dose limits. It is uncertain whether this can be achieved.

(354) Unless a general departure from the quantitative approach to dose limitation can be attained, the simplified system of w_R would create a further problem. The radical simplification of the numerical values would either cause a broad increase of the values of effective dose, or cause a reduction of the values of effective dose for some radiations, such as neutrons. This would cause either an unintended tightening of the limits for a broad class of radiation types, or would generate pressure to offset the change by reducing the annual effective dose limits, which would then apply to all radiations, including photons.

6.5.3. Need for a formal relationship between w_R and $Q(L)$

(355) The concept of w_R with reference to the external field was introduced in order to provide computational simplification. Some simplification was achieved by the introduction of w_R, but due to the by-now greatly simplified computational procedures, this aspect has lost some relevance. In some cases, e.g. in dealing with radiation fields in aviation altitudes, the intended simplification amounts, in fact, to a complication.

(356) The definition in terms of w_R sets the effective dose conceptually apart from the operational quantities H^* and H_p which continue to be defined in terms of $Q(L)$. The lack of equivalence between w_R and $Q(L)$ that depends on the internal field separates 'computational dosimetry', something of a contradictio in adjecto, from dosimetry which involves measurements. The separation is artificial but acceptable in most routine applications; however, it can create difficulties in those critical cases where compliance with the dose limits needs to be assessed with some precision. There is, accordingly, a need to attain a more coherent system of w_R values.

6.6. Proposed convention for neutrons

6.6.1. The intended relationship between w_R and $Q(L)$

(357) When *Publication 60* (ICRP, 1991) introduced w_R and presented a new numerical convention for $Q(L)$, it stated that the two concepts are broadly compatible, but no formal relationship between w_R and $Q(L)$ was identified. However, it was suggested that w_R is largely equivalent to the ambient quality factor q^*, i.e. the mean quality factor at the reference position at depth 10 mm in the ICRU sphere.

(358) Coherence with the concept of the earlier reference quantity, H_E, would have been obtained if w_R had been set equal to the effective quality factor, q_E, in an anthropomorphic phantom (for a chosen standard exposure situation, such as isotropy). However, the numerical data for q_E of neutrons were not available at the time of *Publication 60* (ICRP, 1991) and, accordingly, reference was made to q^*.

(359) When w_R was based on q^*, it was realised that q^* fails to account for the substantial dose contribution in the human body or in the phantom from photons generated by neutrons of energy of about 1 MeV or less, but the resulting imbalance—i.e., substantially larger implied weight factor for the high LET component from low-energy neutrons than for the high LET component from high-energy neutrons—was judged to be tolerable. Now, as accurate values of q_E are available, the imbalance can be removed.

6.6.2. Proposed modification of w_R

(360) The present difficulties could be removed by changing the numerical values of w_R so that they equal q_E for incident neutrons of specified energy. However, this would considerably reduce the current magnitude of the effective dose from a specified fission neutron exposure. Since the current definition has become part of the radiation protection regulations in those countries where the ICRP recommendations have been implemented, the reduction does not appear to be feasible by now. The reduction of the magnitude of the effective dose for a given neutron exposure is also undesirable because it would make the risk estimate for fission neutrons per unit effective dose larger than the nominal risk coefficient currently specified by the ICRP.

(361) The recommended alternative is represented in Fig. 6.1. It is a modification of w_R that largely preserves the current magnitude of the effective dose, while it corrects the spuriously high values of w_R at neutron energies below 1 MeV (see Fig. 4.4). The modified values of w_R exhibit, in the case of neutrons, the same energy dependence as the effective quality factor, q_E, but they are scaled up, roughly by a factor of 1.6, to reach a maximum of 20 at 1 MeV that preserves the current value of w_R at this neutron energy.

(362) The proposed modification retains w_R as an external radiation weighting factor. However, the modified w_R is coherent (for isotropic exposure) with the LET-dependent weighting factor which has, apart from a scaling factor [see Eq(4.7)], the same LET dependence as the current quality factor. The provision of a weighting factor that depends on the LET of the internal field and is essentially equivalent to

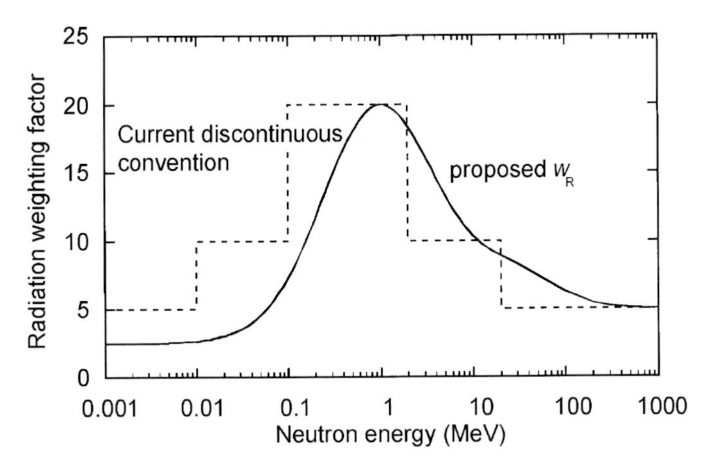

Fig. 6.1. The current convention (ICRP, 1991) for the radiation weighting factor w_R as discontinuous function of the neutron energy (broken line) and the proposed modification (solid line) that equals, apart from a scaling factor, the effective quality factor q_E for the neutrons.

the proposed w_R closes the current gap between computational determinations of the effective dose or H_T and determinations that involve dosimetric and microdosimetric measurements. The LET-dependent weighting factor can be applied not only in measurements, but also in computations—for example, in space or at aviation altitudes—where reference to an LET-dependent weighting factor can be more straightforward than an assessment in terms of the separate radiation components and their still uncertain w_R values. It also provides flexibility in those exceptional situations where w_R values are not defined or are not readily applicable; for example, highly non-uniform exposures to complex radiation fields.

6.6.3. Neutrons at aviation altitudes and in space

(363) Neutrons of very high energy, with a dominant peak at 100 MeV, and their high-energy secondaries produce the high-LET component which amounts to about 25% of the absorbed dose at aviation altitudes and—if $w_R = 5$ for protons is abandoned—to about 80% of the effective dose. Actual values depend, apart from various other factors, strongly on geographic latitude, diurnal variations, and solar activity.

(364) w_R values for neutrons in high altitude and space exposures are included in the proposed convention. $w_R = 6$ for 100 MeV neutrons is a suitable standard value, see Fig. 6.1. However, it must be noted that the w_R values for neutrons at energies of 100 MeV or more may still be subject to considerable change, as new computations become available.

(365) Radiation at aviation altitudes is in an equilibrium that changes little when the radiation is further degraded in the body. Organ-equivalent doses can therefore be obtained by integrating—either through computations or measurements—the

LET-weighted charged particle fluences at a point outside the body. The use of the weighting factor 1.6 $Q(L)-0.6$ [see Eq(4.7)] is, thus, a convenient alternative to the use of w_R.

(366) Radiation protection in space poses specific problems and can require assessments with more than the customary precision. Special regulations and rules may thus be necessary (NCRP, 2000), but the use of the proposed coherent radiation weighting factors will be essential.

6.6.4. Adoption of a continuous dependence of w_R on neutron energy

(367) w_R is currently specified as a step function in neutron energy. A continuous function has been offered as an 'approximation'. Virtually all detailed computations have used the continuous function in order to avoid meaningless discontinuities. Since it is more convenient, with present computing techniques, to use the continuous function, and since discontinuities in the conversion factors can cause practical problems, it is proposed to treat the continuous dependence as the primary convention and to offer the step function as an approximation that is acceptable in most situations.

6.6.5. Continued use of the operational quantities for monitoring

(368) The operational quantities $H^*(d)$ and $H_p(d)$ were introduced for purposes of monitoring, i.e. to substitute for H_E in those cases where no precise assessment is required. In exceptional critical cases, calculations, and possibly measurements in realistic phantoms, could permit improved approximations.

(369) With the current quantity effective dose, i.e. with w_R, but without an equivalent LET-dependent weighting factor, it is difficult or impossible to perform any measurements that can add to the information provided by the operational quantities, $H^*(d)$ or $H_p(d)$. The current w_R and its proposed modification correspond, as stated in Section 6.6.2, to a weighting factor in terms of LET that is larger than $Q(L)$. If reference to $Q(L)$, rather than w_R, were the only difference, the operational quantities would underestimate the effective dose. However, another factor offsets the difference. The operational quantities for penetrating radiations are defined in terms of the comparatively shallow depth of 10 mm of the reference point for absorbed dose. This makes them adequately conservative relative to the effective dose, even though $Q(L)$ is low against w_R. While this balance of two imperfect choices is accidental, it removes the immediate need to change the present definition of $H^*(d)$ and $H_p(d)$.

(370) Measurements with tissue-equivalent proportional counters tend to become increasingly important. They require the specification of the weighting factor in terms of unrestricted LET, L, or its microdosimetric analogue lineal energy, y. With the modified radiation weighting factor, it is possible, at least in principle, to make measurements in realistic phantoms and in terms of the LET-dependent weighting factor that is coherent with w_R. The use of the operational quantities can then be reduced to their original purpose as monitoring quantities.

6.7. Proposed convention for heavy ions

6.7.1. w_R for protons

(371) *Publication 60* (ICRP, 1991) recommended $w_R = 5$ for protons with energy greater than 2 MeV, but did not detail the radiobiological data, or the RBE values, on which the selections of this recommendation were based.

(372) At the time of *Publication 60* (ICRP, 1991), energetic protons were considered to be fairly unimportant to radiation protection, and a very conservative w_R has, thus, been adopted. However, accounting for the radiation quality of protons has become important with regard to the effective doses that can be incurred by crews of aircraft flying at high altitudes. With $w_R = 5$, the protons can contribute a major part of the effective dose.

(373) An assessment in terms of LET shows that energetic protons at aviation altitudes will, even at the largest depth in the human body, have mean energies close to 100 MeV. The resulting mean quality factor for the protons will, therefore, not usually be larger than about 1.15. The effective quality factor can, due to the secondary particles from nuclear interactions within the body, be larger, and a typical value for protons between 1 and 2 GeV is 1.6. The corresponding w_R is 2 [see Eq(4.7)]. It is proposed to adopt this value for cosmic-ray protons.

(374) Exposures in space may require more detailed accounting, and lower energy protons need to be accounted for in the free and, especially, the trapped radiation (NCRP, 2000). A somewhat lower w_R value (see Fig. 4.7) can then be used if precise values are sought. An alternative is the more direct assessment in terms of the LET-dependent weighting factor.

6.7.2. Heavier ions

(375) Heavy ions have not, in the past, been seen as a major problem in radiation protection, and α particles, fission fragments, and heavy nuclei have in summary fashion been assigned a w_R of 20.

(376) Internal emitters must be treated as a separate case because their RBE depends not merely on radiation quality, but also, and particularly for α rays with their short ranges, on their distribution within the tissues or organs. It is, accordingly, unlikely that a single w_R should adequately represent the RBE_M for different α emitters and for different organs, and this is specifically so because of the remaining uncertainties for leukaemia and other blood dyscrasias. The current w_R of 20 for α rays can, thus, serve as a guideline, while for specific situations, such as exposure to radon and its progeny, or the incorporation of ^{224}Ra, ^{226}Ra, thorium, and plutonium, more meaningful weighting factors need to be derived. This can be achieved either in terms of specific assumptions on critical target cells and the resulting dosimetric models, or it is done on the basis of epidemiological information. Specifically, it follows from these considerations that a convention in terms of LET should not include the case of α rays.

(377) External exposure to heavy charged particles is an issue of interest to radiation protection in view of the increased attention to exposures in aviation

altitudes and in space, but also to potential exposures near heavy ion accelerators. The highly simplified convention $w_R = 20$ is, therefore, adequate under conventional circumstances, but it is unsuitable when a realistic assessment is required—this must be done in terms of $Q(L)$ rather than w_R. A main consideration here is that the determination of suitable w_R values—and, thereby, the determination of the effective dose and the equivalent organ doses—ought to be coherent with the system for neutrons. If the suggested modification for w_R for neutrons is adopted, coherence can be achieved in terms of the LET-dependent weighting factor that is implicitly equivalent to w_R. In effect, this amounts to adopting values of w_R for heavy ions that correspond to $Q(L)$, but include the same scaling factor 1.6 [see Eq(4.7)] that is implied in the proposed w_R convention for neutrons (see Section 6.6.2).

6.8. Radiation weighting for deterministic effects

(378) It is assumed that deterministic effects are not caused by ionising radiation below certain dose thresholds that tend to be larger, and often substantially larger, for low-LET radiation than a few hundred mGy. Accordingly, such effects are taken to be of no concern in most radiation protection situations where exposures are distinctly lower.

(379) There is, on the other hand, increasing attention on the substantial exposures associated with prolonged space missions and an awareness of the current lack of knowledge on the precise effect of such exposures (Fry, 2001). However, even apart from these new developments, it has been recognised that there can be special circumstances where higher doses to an organ or a tissue are possible and where the limitation of the effective dose alone cannot rule out such exposures. It has, therefore, been necessary to adopt as additional constraints, an annual equivalent dose limit of 0.15 Sv to the lens of the eye, 0.5 Sv to the skin (averaged over any 1 cm^2 area), and, equally, 0.5 Sv to the hands and feet.

(380) Deterministic effects are taken to be the consequence of radiation-induced cell killing and the resulting depletion of critical tissues. Lens opacification is the important exception because it does not reflect cell killing but cell damage which causes an abnormal differentiation and thus opacities, which, if sufficiently large, may cause visual impairment.

(381) Deterministic effects that are due to cell killing have comparatively low values of RBE for high-LET radiation. Lens opacification exhibits substantially higher values. As far as radiation weighting is concerned, the two types of deterministic effects must, therefore, be clearly distinguished.

6.8.1. Deterministic effects due to cell killing

(382) *Publication 58* (ICRP, 1990) equated the RBE for deterministic effects with the low-dose-limit RBE_m for cell killing. This was a conservative approach that was bound to provide somewhat larger values than the RBEs that actually apply vs the annual occupational dose limit of 1.5 Gy of low-LET radiation, or against the thresholds that are known to be of the order of a few Gy for low-LET radiation. The approach served as adequate assurance that the relevant RBE values for the

deterministic effects are always lower than w_R values for stochastic effects. It was, thus, safely inferred that the 0.15 Sv occupational dose limit provided at least as much protection against high-LET radiation as against low-LET radiation.

(383) The conclusion is, accordingly, that it is sufficiently conservative to use w_R with regard to deterministic effects. The specification in terms of equivalent dose may, on the other hand, be overconservative in special situations where high-LET radiation is the critical factor and where it predominantly exposes a single tissue, such as the skin. It will then be more appropriate to express the dose limit with regard to the deterministic effect in terms of the absorbed dose weighted by RBE_m.

(384) For an even more realistic specification, it is desirable to invoke the RBE for the deterministic effect against the threshold dose in terms of low-LET radiation. This RBE must either be derived from RBE_m and the assumed crossover dose for cell killing [Section 5.2.2; Eq(5.6)] or, especially for those effects that are subject to other modifying factors in addition to cell killing, it must be obtained from experimental studies or clinical observations.

(385) To avoid confusion, it is proposed to designate the unit Gy-Eq whenever an RBE-weighted absorbed dose is used.

6.8.2. Lens opacification

(386) Lens opacification was attributed to accumulated cell damage and was classified as a deterministic effect. However, it reflects an abnormal differentiation rather than cell killing, which sets it apart from other deterministic effects and explains the much higher RBE values that have been seen for this effect.

(387) For low-LET radiation, it has been thought that there was a threshold, both for slight opacifications that do not impair visual acuity and for substantial opacifications with clinical relevance. For the latter, the low-LET dose threshold has, on the basis of clinical studies, been assumed to be at least 2 Gy from an acute exposure.

(388) It is less likely that there is a threshold for slight opacifications with regard to high-LET radiation exposure. In fact, the RBE of 0.4 MeV neutrons against 0.15 Gy of x rays appears to exceed 100 according to the study of Bateman et al. (1972) and 300 according to Worgul et al. (1996). However, the possibility of a non-threshold dose–response relationship has been suggested. If there is a threshold, the study of the tinea capitis patients suggests that it may be as low as 0.5 Gy (Albert et al., 1968). There is considerably less confidence than in the past that a threshold of significant magnitude exists, even for low-LET radiation. If, as has been suggested, small opacities may progress with time, it would be wise to re-examine the recommendations for dose limits. If the damage to the lens is cumulative, the current annual limit of 150 mSv could result in doses to the lens, over a working life-time, that appear too high. The lack of data for humans exposed to high-LET radiations and concern how the experimental data should be applied makes it difficult to recommend weighting factors to be used in setting limits for protection of the lens. It is suggested that a task group examine all the recent data on the lens and make proposals to the Commission on dose limits for low-LET and other radiation qualities.

7. REFERENCES

Albert, R., Omran, A., Brauer, E., et al. (1968) Follow-up study of patients treated by x-ray epilation for tinea capitis. II. Results of clinical and laboratory examinations. Arch. Environ. Health 17, 919–934.

Alpen, E.L., Powers-Risius, P., Curtis, S.B., et al. (1994) Fluence based relative biological effectiveness for charged particle carcinogenesis in mouse Harderian gland. Adv. Space Res. 14, 573–581.

Andersson, M., Carstensen, B., Visfeldt, J. (1993) Leukaemia and other related hematological disorders among Danish patients exposed to Thorotrast. Radiat. Res. 134, 224–233.

Bateman, J.L., Rossi, H.H., Kellerer, A.M., et al. (1972) Dose-dependence of fast neutron RBE for lens opacification in mice. Radiat. Res. 51, 381–390.

Bauchinger, M., Schmid, E., Streng, S., et al. (1983) Quantitative analysis of the chromosome damage at first division of human lymphocytes after ^{60}Co-γ-irradiation. Radiat. Env. Biophys. 22, 225–229.

Baverstock, K.F., Charlton, D.E. (Eds.) (1988) *DNA Damage by Auger Emitters.* Report on a Workshop on 17 July 1987. Taylor and Francis, London, UK.

Belkacemi, Y., Ozsahin, M., Pene, F., et al. (1996) Cataractogenesis after total body irradiation. Int. J. Radiat. Oncol. Biol. Phys. 35, 53–60.

Blakely, E.A., Kronenberg, A. (1998) Heavy-ion radiobiology: new approaches to delineate mechanisms underlying enhanced biological effectiveness. Radiat. Res. 150, S126–S145.

Boice, J.D. Jr. (1993) Leukaemia risk in thorotrast patients. Radiat. Res. 136, 301–302.

Bond, V.P., Meinhold, C.B., Rossi, H.H. (1978) Low-dose RBE and Q for x ray compared to gamma-ray radiations. Health Phys. 34, 433–438.

Borek, C., Hall, E.J., Rossi, H.H. (1978) Malignant transformation in cultured hamster embryo cells produced by x rays, 430 keV monoenergetic neutrons and heavy ions. Cancer Res. 38, 2997–3005.

Borek, C., Hall, E.J., Zaider, M. (1983) x rays may be twice as potent as gamma rays for malignant transformation at low doses. Nature 301, 156–158.

Bozkurt, A., Chao, T.C., Xu, X.G. (2000) Fluence-to-dose conversion coefficients from monoenergetic neutrons below 20 MeV based on the Vip-Man anatomical model. Phys. Med. Biol. 45, 3059–3079.

Bozkurt, A., Chao, T.C., Xu, X.G.Fluence-to-dose conversion coefficients based on the Vip-Man anatomical model and MCNPX code for monoenergetic neutrons above 20 MeV. Health Physics 81, 184–202.

Breckon, G., Cox, R. (1990) Alpha particle leukaemogenesis. Lancet 1, 656–657.

Brenner, D.J., Medvedovsky, C., Huang, Y., et al. (1993) Accelerated heavy particles and the lens. VIII. Comparisons between the effects of acute low doses of iron ions (190 keV/µm) and argon ions (88 keV/µm). Radiat. Res. 133, 198–203.

Brenner, D.J., Miller, R.C., Huang, Y., et al. (1995) The biological effectiveness of radon-progeny alpha particles. III. Quality factors. Radiat. Res. 143, 61–69.

Brenner, D.J., Sawant, S.G., Hande, M.P., et al. (2002) Routine screening mammography: how important is the radiation-risk side of the benefit-risk equation? Int. J. Radiat. Biol. 78, 1065–1067.

Broerse, J.J., Hennen, L.A., van Zweiten, M.J. (1985) Radiation carcinogenesis in experimental animals and its implications for radiation protection. Int. J. Radiat. Biol. 48, 167–187.

Broerse, J.J., van Bekkum, D.W., Zoetelief, J., et al. (1991) Relative biological effectiveness for neutron carcinogenesis in monkeys and rats. Radiat. Res. 128 (Suppl.), S128–S135.

Brooks, A.L. (1975) Chromosome damage in liver cells from low dose rate alpha, beta and gamma irradiation: derivation of RBE. Science 190, 1090–1092.

Burchall, P.R.J., James, A.C. (1994) Uncertainty analysis of the effective dose per unit exposure from radon progeny and implications for ICRP risk weighting factors. Radiat. Prot. Dosim. 53, 133–140.

Burns, F.J., Albert, R.E. (1981) Dose response for rat skin tumors induced by single and split doses of argon ions. In: Pirruciello, M.C., Tobias, C.A. (Eds.), *Biological and Medical Research with Accelerated Heavy Ions at the Bevalac, Berkeley*. University of California Press, Berkeley, USA, pp. 223–235.

Burns, F.J., Hosselet, S., Garte, S.J. (1989) Extrapolations of rat skin tumor incidence: dose, fractionation and linear energy transfer. In: Baverstock, K.F., Stather, J.W. (Eds.), *Low Dose Radiation Biological Bases of Risk Assessment.* Taylor and Francis, London, UK, pp. 571–584.

Burns, F.J., Zhao, P., Xu, G., et al. (2001) Fibroma induction in rat skin following single or multiple doses of 1.0 GeV/nucleon ^{56}Fe ions from the Brookhaven Alternating Gradient Synchrotron (AGS). Phys. Med. 17 (Suppl.), 194–195.

Carnes, B.A., Grahn, D., Thomson, J.F. (1989) Dose–response modelling of the life shortening in a retrospective analysis of the combined data from the JANUS program at Argonne National Laboratory. Radiat. Res. 119, 39–56.

Chan, G.L., Little, J.B. (1986) Neoplastic transformation *in vitro*. In: Upton, A.C., Albert, R.E., Burns, F.E., et al. (Eds.), *Radiation Carcinogenesis*. Elsevier, Amsterdam, The Netherlands, pp. 108–136.

Charlton, D.E. (1988) Calculation of single and double strand DNA breakage from incorporated ^{125}I. In: Baverstock, K.F., Charlton, D.E. (Eds.), *DNA Damage by Auger Emitters*. Taylor and Francis, London, UK, pp. 89–100.

CIRRPC (1995) Committee on Interagency Radiation Research and Policy Coordination. *Science Report No. 10. ORAU 95/F-29*. Office of Science and Technology Policy, Executive Office of the President, Washington, USA.

Clapp, N.K., Darden Jr., E.B., Jernigan, M.C. (1974) Relative effects of whole-body sublethal doses of 60-MeV protons and 300 kVp x rays on disease incidences in 1 RF mice. Radiat. Res. 57, 158–186.

Covelli, V., Coppola, M., Di Majo, V., et al. (1989) Tumor induction and life shortening in BC3F1 female mice at low doses of fast neutrons and x rays. Radiat. Res. 113, 362–374.

Cox, R., Thacker, J., Goodhead, D.T., et al. (1977) Mutation and inactivation of cultured mammalian cells by various ionizing radiations. Nature 267, 425–427.

Cucinotta, F.A., Manuel, F.K., Jane, J., et al. (2001) Space radiation and cataracts in astronauts. Radiat. Res. 156, 460–466.

Dalrymple, G.V., Lindsay, I.R., Mitchell, J.C., et al. (1991) Review of the USAF/NASA Proton Bioeffects Project: rationale and acute effects. Radiat. Res. 126, 117–119.

Di Majo, V., Coppola, M., Rebessi, S., et al. (1990) Age-related susceptibility of mouse liver to induction of tumors by neutrons. Radiat. Res. 124, 227–234.

Di Majo, V., Coppola, M., Rebessi, S., et al. (1996) The influence of sex on life shortening and tumor incidence in CBA/Cne mice exposed to x rays and fission neutrons. Radiat. Res. 146, 181–187.

Di Paola, M., Coppola, M., Baarli, T., et al. (1980) Biological responses to various neutron energies from 1 to 600 MeV. A II. Lens opacification in mice. Radiat. Res. 84, 453–461.

Dietze, G., Siebert, B.R.L. (1994) Photon and neutron dose contributions and mean quality factors phantoms of different size irradiated by monoenergetic neutrons. Radiat. Res. 140, 130–133.

Dvorak, R., Kliauga, P. (1978) Microdosimetric measurements of ionization by monoenergetic photons. Radiat. Res. 73, 1–20.

Edwards, A.A., Purrott, R.J., Prosser, J.S., et al. (1980) The induction of chromosome aberrations in human lymphocytes by alpha-radiation. Int. J. Radiat. Biol. Relat. Stud. Phys. Chem. Med. 38, 83–91.

Edwards, A.A., Lloyd, D.C., Purrott, R.J, et al. (1982) The dependence of chromosome aberration yields on dose rate and radiation quality. In: *Research and Development Report, 1979-1981, R&D 4*. National Radiological Protection Board, Chilton, Oxon, UK.

Edwards, A.A., Lloyd, D.C., Prosser, J.S. (1985) Induction of chromosome aberrations in human lymphocytes by accelerated charged particles. Radiat. Prot. Dosim. 13, 205–209.

Edwards, A.A., Lloyd, D.C., Prosser, J.S., et al. (1986) Chromosome aberrations induced in human lymphocytes by 8.7 MeV protons and 23.5 helium-3 ions. Int. J. Radiat. Biol. 50, 137–145.

Edwards, A.A., Moiseenko, V., Nikjoo, H. (1994) Modelling of DNA breaks and the formation of chromosome aberrations. Int. J. Radiat. Biol. 66, 633–637.

Edwards, A.A. (1997) The use of chromosomal aberrations in human lymphocytes for biological dosimetry. Radiat. Res. 148 (Suppl.), S39–S44.

Edwards, A.A. (1999) Neutron RBE values and their relationship to judgements in radiological protection. J. Radiol. Prot. 19, 93–105.

Edwards, A.A. (2001) RBE of radiations in space and the implications for space travel. Phys. Med. 17 (Suppl.), 147–152.

Engels, H., Wambersie, A. (1998) Relative biological effectiveness of neutrons for cancer induction and other late effects: a review of radiological data. Recent Results Cancer Res. 150, 54–87.

Evans, R.D. (1966) The effect of skeletally deposited alpha-ray emitters in man. Br. J. Radiol. 468, 881–895.

Evans, R.D. (1980) Radium poisoning: a review of the present knowledge. Health Phys. 38, 899–905.

Failla, G., Henshaw, P. (1931) The relative biological effectiveness of x rays and gamma rays. Radiology 17, 1–43.

Fedorenko, B.S., Abrosimova, A.N., Smirnova, O.A. (1995) The effect of high-energy accelerated particles on the crystalline lens of laboratory animals. Phys. Part. Nucl. 26, 573–588.

Ford, J., Terzaghi, M. (1993) Effects of ^{210}Pu alpha particles on survival and preoplastic transformation of primary rat tracheal epithelial cells irradiated while in suspension or in intact tissue. Radiat. Res. 136, 89–96.

Frankenberg, D., Kelnhofer, K., Bar, K., et al. (2002) Enhanced neoplastic transformation by mammography x rays relative to 200 kVp x rays: indication for a strong dependence on photon energy of the RBE_M for various end points. Radiat. Res. 157, 99–105. Erratum in Radiat. Res. 2002, 158, 126.

Fry, R.J., Powers-Risius, P., Alpen, E.L., et al. (1985) High-LET radiation carcinogenesis. Radiat. Res. 104 (Suppl.), S188–S195.

Fry, S.A. (1998) Studies of U.S. radium dial workers: an epidemiological classic. Radiat. Res. 150 (Suppl.), S21–S29.

Fry, R.J. (2001) Deterministic effects. Health Phys. 80, 338–343.

Geard, C.R. (1985) Chromosomal aberration production by track segment charged particles as a function of linear energy transfer. Radiat. Prot. Dosim. 13, 199–204.

Gerwick, L.E., Kozin, S.V. (1999) Relative biological effectiveness of proton beams in clinical therapy. Radiother. Oncol. 50, 135–142.

Gössner, W. (1999) Pathology of radium-induced bone tumors: new aspects of histopathology and histogenesis. Radiat. Res. 152 (Suppl.), S12–S15.

Gössner, W., Masse, R., Stather, J.W. (2000) Cells at risk for dosimetric modelling relevant to bone tumor induction. Radiat. Prot. Dosim. 92, 209–213.

Gössner, W. (2003) Target cells in internal dosimetry. In*: Proceedings: Workshop on Internal Dosimetry of Radionuclides, Oxford, September 2002*. Radiat. Prot. Dosim., 105, 39–42.

Grahn, D., Lombard, L.S., Carnes, B.A. (1992) The comparative tumorigenic effects of fission neutrons and cobalt-60 gamma rays in the B6CF1 mouse. Radiat. Res. 129, 19–36.

Grogan, H.A., Sinclair, W.K., Voilleque, P.G. (2001) Risks of fatal cancer from inhalation of ^{239,240}plutonium by humans: a combined four-method approach with uncertainty evaluation. Health Phys. 80, 447–461.

Gueulette, J., Gregoire, V., Octave-Prignot, M., et al. (1996) Measurements of radiological effectiveness in the 85 MeV proton beam produced at the cyclotron CYCLONE of Louvain-La Neuve, Belgium. Radiat. Res. 145, 70–74.

Hall, E.J., Rossi, H.H., Zaider, M., et al. (1982) The role of neutrons in cell transformation research. II Experimental. In: Broerse, J.J., Gerber, G.B. (Eds.), *Neutron Carcinogenesis. CEC Report Eur 8084*. Commission of the European Communities, Luxembourg, pp. 381–396.

Han, A., Elkind, M.M. (1979) Transformation of mouse C3H 10 T 1/2 cells by single and fractionated doses of x rays and fission neutrons. Cancer Res. 39, 123–130.

Hill, C.F., Han, A., Elkind, M.M. (1984) Fission-spectrum neutrons at a low dose rate enhance neoplastic transformation in the linear low dose region (0–10 cGy). Int. J. Radiat. Biol. 46, 11–15.

Hill, C., Carnes, B.A., Han, A., et al. (1985) Neoplastic transformation is enhanced by multiple low doses of fission-spectrum neutrons. Radiat. Res. 102, 404–410.

IARC (2000) World Health Organization, International Agency for Research on Cancer. *IARC Monographs on the Evaluation of Carcinogenic Risks to Humans. Vol 75. Ionizing Radiation, Part 1: X- and Gamma (γ)-Radiation and Neutrons*. IARC, Lyon, France.

IARC (2001) World Health Organization, International Agency for Research on Cancer. *IARC Monographs on the Evaluation of Carcinogenic Risks to Humans. Vol 78. Ionizing Radiation, Part 2: Some Internally Deposited Radionuclides*. IARC, Lyon, France.

ICRP (1951) Recommendations of the International Commission on Radiological Protection and of the

International Commission on Radiological Units (NBS 47, 1950). Revised by the International Commission on Radiological Protection at the Sixth International Congress of Radiology, London, UK, 1950. Brit. J. Radiol. 24, 46–53.

ICRP (1955) Recommendations of the International Commission on Radiological Protection (Revised 1954). Br. J. Radiol. (Suppl. 6).

ICRP (1977) Recommendations of the International Commission on Radiological Protection. ICRP Publication 26, Ann. ICRP 1 (3).

ICRP (1984) Nonstochastic effects of ionizing radiation. ICRP Publication 41, Ann. ICRP 14 (3).

ICRP (1990) RBE for deterministic effects. ICRP Publication 58, Ann. ICRP 20 (4).

ICRP (1991) Recommendations of the International Commission on Radiological Protection. ICRP Publication 60, Ann. ICRP 21 (1–3).

ICRP (1996) Conversion coefficients for use in radiological protection against external radiation. ICRP Publication 74, Ann. ICRP 26 (3/4).

ICRU (1959) *Report 9. Report of the International Commission on Radiological Unit and Measurements.* ICRU, Washington, DC. (NBS 78, 1961).

ICRU, ICRP (1963) Report of the RBE Committee of the International Commissions on Radiological Protection and on Radiological Units. Health Phys. 9, 357–384.

ICRU (1970) *Linear Energy Transfer, ICRU Report 16.* International Commission on Radiation Units and Measurements, Bethesda, Maryland, USA.

ICRU (1980) *Radiation Quantities and Units, ICRU Report 33.* International Commission on Radiation Units and Measurements, Bethesda, Maryland, USA.

ICRU (1983) *Microdosimetry, ICRU Report 36.* International Commission on Radiation Units and Measurements, Bethesda, Maryland, USA.

ICRU (1985) *Determination of Dose Equivalents Resulting from External Radiation Source, ICRU Report 39.* International Commission on Radiation Units and Measurements, Bethesda, Maryland, USA.

ICRU (1986) *The Quality Factor in Radiation Protection, ICRU Report 40.* International Commission on Radiation Units and Measurements, Bethesda, Maryland, USA.

ICRU (1993a) *Stopping Powers and Ranges of Protons and Alpha Particles with Data Disk, ICRU Report 49.* International Commission on Radiation Units and Measurements, Bethesda, Maryland, USA.

ICRU (1993b) *Quantities and Units in Radiation Protection Dosimetry.* International Commission on Radiation Units and Measurements, Bethesda, MD.

Joiner, M.C., Lambin, P., Malaise, E.P., et al. (1996) Hypersensitivity to very-low single radiation doses: its relationship to the adaptive response and induced radioresistance. Mutat. Res. 358, 171–183.

Kassis, A.I., Fayad, F., Kinsey, B.M., et al. (1987) Radiotoxicity of ^{125}I in mammalian cells. Radiat. Res. 111, 305–318.

Kellerer, A.M., Rossi, H.H. (1972) The theory of dual radiation action. Curr. Topics Radiat. Res. Quart. 8, 85–158.

Kellerer, A.M., Rossi, H. (1982) Biophysical aspects of radiation carcinogenesis. In: Becker, F.F. (Ed.), *Cancer. A Comprehensive Treatise, 2nd edn.* Plenum Press, New York, USA, pp. 569–616.

Kellerer, A.M., Hahn, K. (1988a) Considerations on a revision of the quality factor. Radiat. Res. 114, 480–488.

Kellerer, A.M., Hahn, K. (1988b) The quality factor for neutrons in radiation protection: physical parameters. Radiat. Protect. Dosim. 23, 73–78.

Kellerer, A.M., Walsh, L. (2001) Risk estimation for fast neutrons with regard to solid cancer. Radiat. Res. 156, 708–717.

Kellerer, A.M., Walsh, L. (2002) Solid cancer risk coefficient for fast neutrons in terms of effective dose. Radiat. Res. 158, 61–68.

Kellerer, A.M. (2002) Electron spectra and the RBE of x rays. Radiat. Res. 158, 13–22.

Kiefer, J., Schmidt, P., Koch, S. (2001) Mutations in mammalian cells induced by heavy charged particles: an indicator for risk assessment in space. Radiat. Res. 156, 607–611.

Kocher, D.C. (2001) Comments on the radioepidemiological tables, and the supporting interactive computer program (IREP), developed under the EEPICA by NIOSH. http://www.cdc.gov/niosh/ocas/pdfs/.

Kocher, D.C., Apostoaei, A.I., Hoffman, O. (2002) Proposed revision to RBE factors. http://www.cdc.gov/niosh/ocas/pdfs/ireprevf.pdf.

Koshurnikova, N.A., Gilbert, E.S., Shilnikova, N.S., et al. (2002) Studies on the Mayak nuclear workers: health effects. Radiat. Environ. Biophys. 41, 29–31.

Kraft, G. (1987) Radiobiological effects of very heavy ions: Inactivation, chromosomal aberration and strand breaks. Nuclear Sci. Application 3, 1–28.

Kreisheimer, M., Koshurnikova, N.A., Nekolla, E., et al. (2000) Lung cancer mortality among male nuclear workers of the Mayak facilities in the former Soviet Union. Radiat. Res. 154, 3–11.

Lafuma, J., Chemelevsky, D., Chameaud, J., et al. (1989) Lung carcinomas in Sprague-Dawley rats after exposure to low doses of radon daughters, fission neutrons or γ rays. Radiat. Res. 118, 230–245.

Land, C., Gilbert, E., Smith, J., et al. (2002) Report of the NCI-CDC Working Group to revise the 1985 NIH radioepidemiological tables: Overview. Health Phys. 82 (Suppl.), S187–S188.

Lea, D.E. (1946) *Actions of Radiations on Living Cells, 1st edn.* Macmillan, New York, USA [2nd edn., University Press, Cambridge (1956)].

Leuthold, G., Mares, V., Schraube, H. (1992) Calculation of the neutron ambient dose equivalent on the basis of the ICRP revised quality factors. Radiat. Prot. Dosim. 40, 77–84.

Leuthold, G., Mares, V., Schraube, H. (1997) Monte-Carlo calculations of dose equivalents for neutrons in anthropomorphic phantoms using the ICRP 60 recommendations and the stopping power data of ICRU 49. GSF Report.

Lindborg, L. (1976) Microdosimetry measurements in beam of high energy photons and electrons: technique and results. In: Booz, J., Ebert, H.G., Smith, B.G.R. (Eds.), *Proceedings of the 5th Symposium on Microdosimetry, Report No. EUR 5452.* Commission of the European Communities, Luxembourg.

Lloyd, E.L., Gemmell, H.A., Henning, C.B., et al. (1979) Transformation of mammalian cells by alpha particles. Int. J. Radiat. Biol. 36, 467–478.

Lloyd, R.D., Miller, S.C., Taylor, G.N., et al. (1994) Relative effectiveness of ^{239}Pu and some other internal emitters for bone cancer induction in beagles. Health Phys. 67, 346–353.

Lundgren, D.L., Haley, P.J., Hahn, F.H.J., et al. (1995) Pulmonary carcinogenicity of repeated inhalation exposure of rats to aerosols of $^{239}PuO_2$. Radiat. Res. 142, 39–53.

Machinami, N., Ishikawa, Y., Boecker, B.B. (Eds.). (1999). *The International Workshop on the Health Effects of Thorotrast, Radium, Nadon and other Alpha-Emitters.* Radiat. Res. 142, 6 (Suppl.)

Makrigiorgios, G., Adelstein, S.J., Kassis, A.I. (1990) Auger electron emitters: Insights gained from in vitro experiments. Rad. Environ. Biophysics 29, 75–91.

Mares, V., Leuthold, G., Schraube, H. (1997) Organ doses and dose equivalents for neutrons above 20 MeV. Radiat. Protect. Dosim. 70, 391–394.

Martin, S.G., Miller, R.C., Geard, C.R., et al. (1995) The biological effectiveness of radon-progeny alpha particles. IV. Morphological transformation of Syrian hamster embryo cells at low doses. Radiat. Res. 142, 70–77.

Mays, C.W., Finkel, M.P. (1980) RBE of α particles vs. β particles in bone sarcoma induction. In: Proceedings of the 5th International Congress of IRPA, Vol. 11. Israel Health Physics Society, Jerusalem, Israel, p. 41.

Mays, C.W., Spiess, H. (1984) Bone sarcomas in patients given radium-224. In: Boice Jr., J.D., Fraumeni Jr., J.F. (Eds.), *Radiation Carcinogenesis: Epidemiology and Biological Significance.* Raven Press, New York, USA, pp. 241–252.

Mays, C.W., Taylor, G.N., Lloyd, R.D. (1986) Toxicity ratios: their use and abuse in predicting risk from induced cancer. In: Thompson, R.C., Mahaffy, J.A. (Eds.), *Life-span Radiation Effects Studies in Animals: What Can They Tell Us? CONF-830951. Proceedings of the 22nd Hanford Life Sciences Symposium, September 1983.* Office of Scientific and Technical Information, Springfield, Virginia, USA.

Merriam Jr., G.R., Focht, E.F. (1957) A clinical study of radiation cataracts and the relationship to dose. Am. J. Roent. Radium Ther. Nuclear Med. LXVII, 759–785.

Merriam, G.R., Focht, E.F. (1962) A clinical and experimental study of the effect of single and divided doses of radiation on cataract production. Trans. Am. Ophtalmol. Soc. 60, 35–52.

Miller, R.C., Hall, E.J. (1991) Oncogenic transformation of C3H 10T1/2 cells by acute and protracted exposures to monoenergetic neutrons. Radiat. Res. 128 (Suppl.), S60–S64.

Miller, R.C., Marino, S.A., Napoli, J., et al. (2000) Oncogenic transformation in C3H10T1/2 cells by low-energy neutrons. Int. J. Radiat. Biol. 76, 327–333.

Mori, T., Kido, C., Fukutomi, K., et al. (1999) Summary of entire Japanese thorotrast follow-up study: updated 1998. Radiat. Res. 152 (Suppl.), S84–S87.

Morstin, K., Bond, V.P., Baum, J.W. (1989) Probabilistic approach to obtain hit-size effectiveness functions which relate microdosimetry and radiobiology. Radiat. Res. 120, 383–402.

Muirhead, C.R., Cox, R., Stather, J.W., et al. (1997) Relative biological effectiveness: Alpha particles. Documents of the NRPB 4, 129–136.

NAS (1999) National Academy of Sciences. *Health Effects of Exposure to Radon. BEIR VI.* National Academy Press, Washington, USA.

NCRP (1980) *Influence of Dose and its Distribution in Time on Dose-Response Relationships for Low-LET Radiations. NCRP Report No. 64.* National Council on Radiation Protection and Measurements, Bethesda, Maryland, USA.

NCRP (1990) *The Relative Biological Effectiveness of Radiations of Different Quality. NCRP Report 104.* National Council on Radiation Protection and Measurements, Bethesda, Maryland, USA.

NCRP (1991) *Some Aspects of Strontium Radiobiology. NCRP Report 110.* National Council on Radiation Protection and Measurements, Bethesda, Maryland, USA.

NCRP (1993) *Limitation of Exposure to Ionizing Radiation. NCRP Report 116.* National Council on Radiation Protection and Measurements, Bethesda, Maryland, USA.

NCRP (1997) *Uncertainties in Fatal Cancer Risk Estimates Used in Radiation Protection. NCRP Report No. 126.* National Council on Radiation Protection and Measurements, Bethesda, Maryland, USA.

NCRP (2000) *Radiation Protection Guidance for Activities in Low-Earth Orbit. NCRP Report No. 132.* National Council on Radiation Protection and Measurements, Bethesda, Maryland, USA.

Neary, G.J., Munson, R.J., Mole, R.H. (1957) *Chronic Radiation Hazards: An Experimental Study with Fast Neutrons.* Pergamon Press, New York, USA.

Nekolla, E.A., Kreisheimer, M., Kellerer, A.M., et al. (2000) Induction of malignant bone tumors in radium-224 patients: risk estimates based on the improved dosimetry. Radiat. Res. 153, 93–103.

Niemer-Tucker, M.M., Sterk, C.C., Dewolf-Roundaal, D., et al. (1999) Late ophthalomogical complications after total body irradiation in non-human primates. Int. J. Radiat. Biol. 75, 465–472.

NIH (1985) *Publication 85-2748: Report of the NIH Ad Hoc Working Group to Develop Radioepidemiological Tables.* National Institutes of Health, US Department of Health and Human Services, Washington, DC.

NRPB (1997) *Relative Biological Effectiveness of Neutrons for Stochastic Effects. Documents of the NRPB 8 (2).* National Radiological Protection Board, Chilton, Didcot, UK.

Obe, G., Johannes, I., Johannes, C., et al. (1997) Chromosomal aberrations in blood lymphocytes of astronauts after long-term space flights. Int. J. Radiat. Biol. 72, 727–734.

O'Sullivan D. (1999) *Study of Radiation Fields and Dosimetry at Aviation Altitudes. Final Report 1996-1999.* EU Contract Number F14P-CT950011. http://www.dias.ie/dias/cosmic/general/Publications/1999/TechReport/DIAS-99-9-1/DIAS-99-9-1.pdf.

Otake, M., Schull, W.J. (1990) Radiation-related posterior lenticular opacities in Hiroshima and Nagasaki atomic bomb survivors based on the DS86 dosimetry system. Radiat. Res. 121, 3–13.

Paganetti, H., Olko, P., Kobus, H., et al. (1997) Calculation of relative biological effectiveness for proton beams using biological weighting functions. Int. J. Radiat. Oncol. Biol. Phys. 37, 719–729.

Paganetti, H., Niemierko, A., Ancukiewicz, M., et al. (2002) Relative biological effectiveness (RBE) values for proton beam therapy. Int. J. Radiat. Oncol. Biol. Phys. 53, 407–421.

Pelliccioni, W. (1998) Radiation weighting factors and high energy radiation. Radiat. Protect. Dosimet. 80, 371–378.

Pierce, D.A., Preston, D.L. (2000) Radiation-related cancer risks at low doses among atomic bomb survivors. Radiat. Res. 154, 178–186.

Preston, D.L., Mattsson, A., Holmberg, E., et al. (2002) Radiation effects on breast cancer risk: a pooled analysis of eight cohorts. Radiat. Res. 158, 220–235.

Preston, D.L., Shimizu, Y., Pierce, D.A., et al. (2003) Studies of mortality of atomic bomb survivors. Report 13 solid cancer and non-cancer disease mortality: 1950–1977. Radiat. Res. 160, 381–407.

Purrott, R.J., Edwards, A.A., Lloyd, D.C., et al. (1980) The induction of chromosome aberrations in human lymphocytes by in vitro irradiation with alpha-particles from plutonium-239. Int. J. Radiat. Biol. Relat. Stud. Phys. Chem. Med. 38, 277–284.

Raabe, O.G., Book, S.A., Parks, N.J. (1983) Lifetime bone cancer dose-response relationships in beagles and people from skeletal burden of ^{226}Ra and ^{90}Sr. Health Phys. 14 (Suppl.), 33–48.

Raju, M.R. (1995) Proton radiobiology, radiosurgery and radiotherapy. Int. Radiat. Biol. 67, 237–259.

Riches, A.C., Herceg, Z., Bryant, P.E., et al. (1997) Radiation-induced transformation of SV 40-immortalized human thyroid epithelial cells by single exposure to plutonium α particles in vitro. Int. J. Radiat. Biol. 72, 515–521.

Ritter, S., Kraft-Weyrather, W., Scholz, M., et al. (1992) Induction of chromosome aberrations in mammalian cells after heavy ion exposure. Adv. Space Res. 12, 119–125.

Rossi, H.H. (1995) Sensible radiation protection. Health Phys. 69, 394–395.

Rossi, H.H., Zaider, M. (1996) *Microdosimetry and its Applications*. Springer, Berlin, Germany.

Roth, J., Brown, N., Catterall, M., et al. (1976) Effects of fast neutrons on the eye. Br. J. Ophthalmol. 60, 236–278.

Sacher, G.A. (1976) Dose, dose rate radiation quality and host factors for radiation-induced life shortening. In: Smith, K. (Ed.), *Aging, Carcinogenesis and Radiation Biology*. Raven Press, New York, USA, pp. 493–517.

Sasaki, M.S., Kobayashi, K., Hieda, K., et al. (1989) Induction of chromosome aberrations in human lymphocytes by monochromatic x rays of quantum energy between 4.8 and 14.6 keV. Int. J. Radiat. Biol. 56, 975–988.

Sasaki, M.S. (1991) Primary damage and fixation of chromosomal DNA as probed by monochromatic soft x rays and low-energy neutrons. In: Fielden, E.M., O'Neil, P. (Eds.), *The Early Effects of Radiation on DNA. NATO ASI Series, Vol. H54*. Springer-Verlag, Berlin, Germany, pp. 369–384.

Sax, K. (1938) Chromosome aberrations induced by x rays. Genetics 23, 494–516.

Schmid, E., Schraube, H., Bauchinger, M. (1998) Chromosome aberration frequencies in human lymphocytes irradiated in a phantom by a mixed beam of fission neutrons and gamma-rays. Int. J. Radiat. Biol. 73, 263–267.

Schmid, E., Regulla, D., Guldbakke, S., et al. (2000) The effectiveness of monoenergetic neutrons at 565 keV in producing dicentric chromosomes in human lymphocytes at low doses. Radiat. Res. 154, 307–312.

Schmid, E., Regulla, D., Guldbakke, S., et al. (2002a) Relative biological effectiveness of 144 keV neutrons in producing dicentric chromosomes in human lymphocytes compared with ^{60}Co gamma rays under head-to-head conditions. Radiat. Res. 157, 453–460.

Schmid, E., Regulla, D., Kramer, H.M., et al. (2002b) The effect of 29 kV x rays on the dose response of chromosome aberrations in human lymphocytes. Radiat. Res. 158, 771–777.

Schmid, E. (2002) Is there reliable experimental evidence for a low-dose RBE of about 4 for mammography x rays relative to 200 kV x rays?. Radiat. Res. 158, 778–781.

Schmid, E., Schlegel, D., Guldbakke, S., et al. (2003) The RBE of nearly mono-energetic neutrons at energies from 36 keV to 14.6 MeV for inducing dicentrics in human lymphocytes. Radiat. Envir. Bioph., in press.

Shellabarger, C., Kellerer, A.M., Rossi, H.H., et al. (1974) Rat mammary carcinogenesis following neutron or x-irradiation. In: *Biological Effects of Neutron Irradiation*. IAEA, Vienna, Austria.

Shellabarger, C.J., Chmelevsky, D., Kellerer, A.M. (1980) Induction of mammary neoplasms in the Sprague-Dawley rat by 430-keV neutrons and x rays. J. Natl. Cancer Inst. 64, 821–832.

Shellabarger, C.J., Chmelevsky, D., Kellerer, A.M., et al. (1982) Induction of mammary neoplasms in the AC1 rat by 430 keV neutrons and x rays and diethylstilbestrol. J. Natl. Cancer Inst. 69, 1135–1146.

Shimizu, Y., Kato, H., Schull, W.J., et al. (1992) Studies of the mortality of A-bomb survivors. 9. Mortality, 1950-1985: Part 3. Noncancer mortality based on the revised doses (DS86). Radiat. Res. 130, 249–266.

Shimuzu, Y., Pierce, D.A., Preston, D.L., et al. (1999) Studies of the mortality of atomic bomb survivors. Report 12, Part II. Noncancer mortality: 1950–1990. Radiat. Res. 152, 374–389.

Shore, R.E., Worgul, B.V. (1999) Overview of the epidemiology of radiation carcinogenesis. In: Junk, A.K., Kundiev, Y., Vitte, P., et al. (Eds.), *Ocular Risk Assessment in Populations Exposed to Environmental Radiation Contamination*. Kluwer Academic Publishers, Dordrecht, The Netherlands, pp. 183–189.

Sinclair, W.K. (1982) Fifty years of neutrons in biology and medicine; the comparative effects of neutrons in biological systems. In: Booz, J., Ebert, H.A. (Eds.), *Proceedings of the 8th Symposium on Microdosimetry*. European Communities, Luxembourg.

Sinclair, W.K. (1985) Experimental RBE values of high LET radiations at low doses and the implications for quality factor assignment. Radiat. Prot. Dosim. 13, 319–326.

Sinclair, W.K. (1990) Quality Factor, concepts and issues. Radiat. Prot. Dosim. 31, 355–359.

Sinclair, W.K. (1996) The present system of quantities and units for radiation protection. Health Phys. 70, 781–786.

Skarsgard, L.D. (1998) Radiobiology with heavy charged particles: a historical review. Phys. Med. 14 (Suppl.), 1–19.

Sparrow, A.H., Underbrink, A.G., Rossi, H.H. (1972) Mutations induced in Tradescantia by small doses of x rays and neutrons: analysis of dose-response curves. Science 176, 916–918.

Spiers, F.W., Lucas, H.F., Rundo, J., et al. (1983) Leukaemia incidence in the U.S. dial workers. Health Phys. 44 (Suppl.), 65–72.

Stehney, A.H. (1995) Health studies of U.S. women radium dial workers. In: Young, J.P., Yalow, R.S. (Eds.), *Radiation and Public Perception: Benefits and Risks*. American Chemical Society, Washington, USA.

Storer, J.B., Mitchell, T.J. (1984) Limiting values for RBE of fission neutrons at low doses for life shortening in mice. Radiat. Res. 97, 396–406.

Storer, J.B., Fry, R.J.M. (1995) On the shapes of the neutron dose-effect curves for radiogenic cancer and life shortening in mice. Radiat. Environ. Biophys. 34, 21–27.

Straume, T. (1996) Risk implications of the neutron discrepancy in the Hiroshima DS86 dosimetry system. Radiat. Prot. Dosim. 67, 9–12.

Takatsuji, T., Sasaki, M.S. (1984) Dose-effect relationship of chromosome aberrations induced by 23 MeV alpha particles in human lymphocytes. Int. J. Radiat. Biol. Relat. Stud. Phys. Chem. Med. 45, 237–243.

Taucher-Scholz, G., Kraft, G. (1999) Influence of radiation quality on the yield of DNA strand breaks for SV40 DNA irradiated in solution. Radiat. Research 151, 595–604.

Taylor, G.N., Mays, C.W., Lloyd, R.D., et al. (1983) Comparitive toxicity of ^{226}Ra, ^{230}Pu, ^{249}Cf and ^{252} in C57 Bl/Do and albino mice. Radiat. Res. 95, 584–601.

Taylor, L. (1984) *The Tripartite Conference on Radiation Protection: Canada, United Kingdom and United States (1949-53). US Department of Energy, Publication NVO-271 (DE 84016028)*. US Department of Energy, Washington, USA.

Terzaghi, M., Ford, J. (1994) Effects of radiation on rat respiratory cells: critical target cell populations and the importance of cell-cell interactions. Adv. Space Res. 14, 565–572.

Testard, J., Ricoul, M., Hoffschir, F., et al. (1996) Radiation-induced chromosome damage in astronauts' lymphocytes. Int. J. Radiat. Biol. 70, 403–411.

Testard, I., Dutrillaux, B., Sabatier, L. (1997) Chromosomal aberrations induced in human lymphocytes by high-LET irradiation. Int. J. Radiat. Biol. 72, 423–433.

Thomas, R.H., Lindell, B. (2001) In radiological protection, the protection quantities should be expressed in terms of measurable physical quantities. Radiat. Prot. Dosim. 94, 287–292.

Thomas, R.H. (2001) The impact of ICRP/ICRU quantities on high energy neutron dosimetry: a review. Radiat. Prot. Dosim. 96, 407–422.

Thomas, R.H., McDonald, J.C., Goldfinch, E.P. (2002) The ICRP and dosimetry: Glasnost Redux. Radiat. Prot. Dosim. 102, 195–200.

Thomson, J.F., Williamson, F.S., Grahn, D. (1983) Life shortening in mice exposed to fission neutrons and gamma rays. III. neutron exposures of 5 and 10 rad. Radiat. Res. 93, 205–209.

Thomson, J.F., Williamson, F.S., Grahn, D. (1985) Life shortening in mice exposed to fission neutrons and γ rays. V. Further studies with single low doses. Radiat. Res. 104, 420–428.

Tobias, C.A., Grigoryev, Y.G. (1975) Ionizing radiation. In: Calvin, M., Gazenko, O.G. (Eds.), *Foundations of Space Biology and Medicine, Vol. 2.* National Aeronautics and Space Administration, Scientific and Technical Information Office, Washington, USA, pp. 473–530.

Ullrich, R.L., Jernigan, M.C., Cosgrove, G.E., et al. (1976) The influence of dose and dose rate on the incidence of neoplastic disease in RFM mice after neutron radiation. Radiat. Res. 68, 115–131.

Ullrich, R.L., Jernigan, M.C., Storer, J.B. (1977) Neutron carcinogenesis. Dose and dose-rate effects in BALB/c mice. Radiat. Res. 72, 487–498.

Ullrich, R.L. (1983) Tumor induction in BALB/c female mice after fission neutrons or gamma irradiation. Radiat. Res. 93, 506–515.

Ullrich, R.L. (1984) Tumor induction in BALB/c mice after fractionated or protracted exposures to fission spectrum neutrons. Radiat. Res. 97, 587–597.

Ullrich, R.L., Preston, R.J. (1987) Myeloid leukaemia in male RFM mice following irradiation with fission spectrum neutrons or gamma rays. Radiat. Res. 109, 165–170.

UNSCEAR (1982) *Ionizing Radiation: Sources and Biological Effects. 1982 Report to the General Assembly, with Annexes.* United Nations Scientific Committee on the Effects of Atomic Radiation. United Nations, New York, USA.

UNSCEAR (1988) *Sources, Effects and Risks of Ionizing Radiation, 1988 Report to the General Assembly, with Annexes.* United Nations Scientific Committee on the Effects of Atomic Radiation. United Nations, New York, USA.

UNSCEAR (1993) *Sources and Effects of Ionizing Radiation, 1993 Report to the General Assembly, with Annexes.* United Nations Scientific Committee on the Effects of Atomic Radiation. United Nations, New York, USA.

UNSCEAR (2000) *Sources and Effects of Ionizing Radiation, Vol. 2, Annex I: Epidemiological Evaluation of Radiation-induced Cancer.* United Nations Scientific Committee on the Effects of Atomic Radiation. United Nations, New York, USA.

Upton, A.C., Randolph, M.L., Conklin, J.W., et al. (1970) Late effects of fast neutrons and gamma-rays in mice as influenced by the dose rate of irradiation: induction of neoplasia. Radiat. Res. 41, 467–491.

Urano, M., Verhey, L.J., Goitein, M., et al. (1984) Relative biological effectiveness of modulated proton beams in various murine tissues. Int. J. Radiat. Oncol. Biol. Phys. 10, 509–514.

van Kaick, G., Dalheimer, A., Hornik, S., et al. (1999) The German thorotrast study: recent results and assessment of risks. Radiat. Res. 152 (Suppl.), S64–S71.

Wolf, C., Lafuma, J., Masse, R., et al. (2000) Neutron RBE for tumors with high lethality in Sprague-Dawley rats. Radiat. Res. 154, 412–420.

Wood, D.H. (1991) Long-term mortality and cancer risk in irradiated Rhesus monkeys. Radiat. Res. 126, 132–140.

Worgul, B.V., Medvedovsky, C., Huang, Y., et al. (1996) Quantitative assessment of the cataractogenic potential of very low doses of neutrons. Radiat. Res. 145, 343–349.

Wouters, B.G., Skarsgard, L.D. (1997) Low-dose radiation sensitivity and induced radioresistance to cell killing in HT-29 cells is distinct from the "adaptive response" and cannot be explained by a subpopulation of sensitive cells. Radiat. Res. 148, 435–442.

Yang, T.C., Craise, L.M., Mei, M.T., et al. (1985) Neoplastic cell transformation by heavy-charged particles. Radiat. Res. 104 (Suppl.), S177–S187.

Yang, T.C., Mei, M., George, K.A., et al. (1996) DNA damage and repair in oncogenic transformation by heavy ion radiation. Adv. Space Res. 18, 149–158.

Zaider, M., Brenner, D.J. (1985) On the microdosimetric definition of quality factors. Radiat. Res. 103, 302–306.

Zirkle, R.E., Marchbank, D.F., Kuck, K.D. (1952) Exponential and sigmoid survival curves resulting from alpha and X-irradiation of Aspergillus spores. J. Cellular Comp. Physiol. 39 (Suppl.), 75–85.